JAMESTOWN'S

Number Power

Robert Mitchell & Donald Prickel

JAMESTOWN PUBLISHERS

a division of NTC/CONTEMPORARY PUBLISHING GROUP
Lincolnwood, Illinois USA

ISBN: 0-8092-2289-2

Published by Jamestown Publishers,
a division of NTC/Contemporary Publishing Group, Inc.,
4255 West Touhy Avenue,
Lincolnwood (Chicago), Illinois 60712-1975 U.S.A.
© 2001 by NTC/Contemporary Publishing Group, Inc.

1 2 3 4 5 6 7 8 9 VH 16 15 14 13 12 11 10 9 8 7 6 5 4 3 2 1

Table of Contents

To the Student vi

Geometry Pretest 1

BUILDING NUMBER POWER 9

Angles

Angles Skills Inventory 10

Introducing Angles 14

Measurement of an Angle 16

Types of Angles 18

Labeling an Angle 20

Measuring an Angle with a Protractor 21

Drawing an Angle with a Protractor 24

Pairs of Angles 26

Pairs of Angles: Applying Your Skills 28

Parallel Lines Cut by a Transversal 30

Parallel Lines Cut by a Transversal: Applying Your Skills 32

Angles Review 34

Triangles

Triangles Skills Inventory 38

Introducing Triangles 42

The Sum of Angles in a Triangle 44

The Sum of Angles in a Triangle: Applying Your Skills 46

Working with Definitions of Triangles 48

Working with Definitions of Triangles: Applying Your Skills 51

Similar Triangles 52

Similar Triangles: Applying Your Skills 60

Squares and Square Roots 62

Right Triangles and the Pythagorean Theorem 64

Right Triangles and the Pythagorean Theorem:
 Applying Your Skills 68
Triangles Review 70

Plane Figures

Plane Figures Skills Inventory 73
Working with Customary Length Units 78
Working with Metric Length Units 80
Addition and Subtraction of Length Units 82
Multiplication and Division of Length Units 84
Working with Length Units: Applying Your Skills 86
Perimeter 88
Recognizing Common Polygons 90
Area 92
Working with Squares 94
Working with Rectangles 98
Working with Triangles 102
Working with Parallelograms and Trapezoids 106
Working with Circles 110
Perimeter and Area: Applying Your Skills 114
Solving Two-Step Area Problems: Part I 118
Solving Two-Step Area Problems: Part II 120
Plane Figures Review 122

Solid Figures

Solid Figures Skills Inventory 125
Recognizing Common Solid Figures 127
Volume 129
Working with Cubes 131
Working with Rectangular Solids 133
Working with Cylinders 135
Working with Cones 136
Volume: Applying Your Skills 137
Solving Two-Step Volume Problems: Part I 139
Solving Two-Step Volume Problems: Part II 141
Solid Figures Review 143

Geometry Posttest A 146

Geometry Posttest B 152

USING NUMBER POWER 159

The Work Triangle 160
Applying the Pythagorean Theorem 162
Tiling a Room 164
Building a Patio 166
Carpeting a Room 168
Yard Work 170
Changing Volume to Capacity 172
Design Work 174
Tessellations 176
Drawing 2-Dimensional Views of 3-Dimensional Objects 177
Precision in Measurement 178
Dimensions of a Sphere 179
Using Geometry in Photography 180
Geometric Illusions 182

Using a Calculator 184

Using Mental Math 185

Using Estimation 186

Formulas 187

Glossary 189

Index 193

To The Student

Welcome to *Geometry*.

Geometry is the study of measurement and relationships of lines, angles, plane (flat) figures, and solid figures. Important topics from geometry include the study of angles and triangles and the study of distance, area, and volume. An understanding of these topics is necessary in many occupations. Also, geometry is a standard section on educational and vocational tests, including GED, college entrance, civil service, and military tests.

This workbook is designed to prepare you for taking a test and for pursuing further education or training that requires basic geometry. The first part of the book, Building Number Power, is divided into four chapters that provide step-by-step instruction in the fundamentals of geometry. Each chapter begins with a skills inventory to help identify geometric skills you work on. Each chapter ends with a skills review to check your progress on newly acquired skills.

The second part of the book, Using Number Power, gives you a chance to apply geometric skills in more detail. These applications are fun and are examples of the use of geometry in everyday life.

Several skills that you will find helpful in your study of geometry are estimation, mental math, and calculator use. These skills are reviewed on pages 184, 185, and 186. Also, for ready reference, a formula summary has been provided on pages 187 and 188, and a glossary is on pages 189–192. The following icons will alert you to problems where estimation, mental math, or calculator use will be especially helpful.

 calculator icon

 estimation icon

 mental math icon

To get the most out of your work, do each problem carefully. Inside the back cover is a chart to help you keep track of your score on each exercise.

Geometry Pretest

This pretest will tell you which chapters of *Geometry* you need to work on and which you already have mastered. Do all the problems that you can. There is no time limit. Use the chart at the end of the test to find the pages where you need to work.

1. What is the measure of the angle formed by the straight sides of each part of the divided circle shown at right?

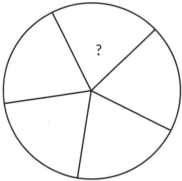

The circle is divided into five equal parts.

2. Draw an acute angle and tell the approximate measure of the angle you draw. Tell how you know your angle is an acute angle.

3. ∠C and ∠D are complementary angles. If the measure of ∠C is 40°, what is the measure of ∠D?

4. A telephone pole has been blown over by a strong wind. The leaning pole makes an acute angle of 47° with the ground. What is the measure of the obtuse angle that the pole makes with the ground?

5. In the drawing at right, label each obtuse angle with its measure. One obtuse angle is labeled as an example. Assume 7th Avenue is parallel to 8th Avenue.

7th Ave.

8th Ave.

Franklin Blvd.

6. What is the measure of ∠2?

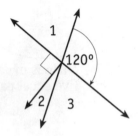

7. What is the measure of ∠E in triangle DEF?

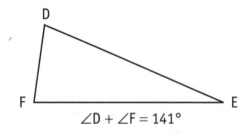

∠D + ∠F = 141°

8. A support cable is attached to a flag pole. The cable makes an angle of 72° with the ground. What angle does the cable make with the pole?

9. Wanda's garden is in the shape of a triangle. What is the measure of the angle that is not labeled?

Wanda's Garden

10. To measure the distance (*d*) across Deer Lake at Rainbow Camp, Luke marked off several distances and drew a map. Using Luke's map, write a proportion that can be used to find the distance (*d*) across Deer Lake.

△ABC and △CDE are similar triangles.

11. Find the square of each number.

 a. $2^2 = $ _____

 b. $4^2 = $ _____

 c. $6^2 = $ _____

 d. $8^2 = $ _____

12. Find each square root.

 a. $\sqrt{9} = $ _____

 b. $\sqrt{16} = $ _____

 c. $\sqrt{49} = $ _____

 d. $\sqrt{81} = $ _____

13. Ami lives 8 miles directly west of Caren. Heather lives 6 miles directly north of Ami. Place the correct name by each house on the map. Then determine the straight-line distance between Caren's house and Heather's house.

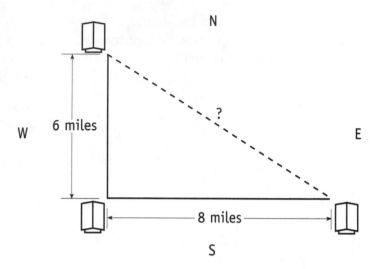

14. The rose garden in Central Park is in the shape of a right triangle. What is the length in yards of the shortest side?

15. How many yards of fencing will it take to enclose a garden that is in the shape of a triangle if the longest side measures 18 yards and the short sides each measure 13 yards?

16. A seamstress cuts a piece of cloth in the shape of a triangle. What is the area of the cloth if the base distance measures 3 yards and the height measures 2 yards?

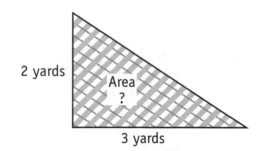

17. How much fence will be needed to enclose a garden that is in the shape of a rectangle if the long side measures 20 yards 2 feet and the short side measures 12 yards 1 foot?

18. Ling cut a piece of plywood in the shape of a parallelogram. The length of the longest side is 6 feet; the length of the shortest side is 4 feet; the height is 3.5 feet. What is the area of the parallelogram?

4 feet 3.5 feet

6 feet

Ling's plywood

19. The running trail in Avery Park is in the shape of a circle. The diameter of the trail is 1 mile. About how many times would Aaron need to run around the trail to run a total distance of 10 miles? ($\pi \approx 3.14$)

Running Trail

diameter = 1 mile

20. The sprinkler shown at right waters in a circular pattern, spraying out water a maximum distance of 9 feet. To the nearest square foot, how much lawn surface does this sprinkler water?

9 feet

21. What is the area of the shaded part of the room pictured at right?

22. a. How many feet are in 1 yard?

b. How many square feet are in 1 square yard?

c. How many cubic feet are in 1 cubic yard?

23. A storage shed with a 10 foot high ceiling is 30 feet long and 24 feet wide. What is the maximum storage volume of the shed?

24. Use the volume formula for a cylinder, $V = \pi r^2 h$, to find the volume of the water storage tower shown at right. Use $\pi \approx \frac{22}{7}$. Express your answer to the nearest cubic foot.

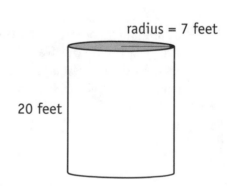

Geometry Pretest Chart

This pretest can be used to identify geometry skills in which you are already proficient. If you miss only 1 question in a section of this test (2 questions in the triangle section), you may not need further study in that chapter. However, before you skip those lessons, we recommend that you complete the review test at the end of that chapter. For example, if you miss 1 question about angles, do the Angles Review (pages 34–36) before beginning Triangles. This longer inventory will be a more precise indicator of your skill level.

PROBLEM NUMBERS	SKILL AREA	PRACTICE PAGES
1	measurement of an angle	16–17
2	types of angles	18–19
3, 6	pairs of angles	26–27
4	pairs of angles: applying skills	28–29
5	parallel lines cut by a transversal	30–31
7	triangle: sum of angles	44–45
8, 9	triangle: applying skills	46–47
10	similar triangles	52–59
11, 12	squares and square roots	62–63
13, 14	Pythagorean theorem	64–67
15	triangle: perimeter	88–89, 102–103
16	triangle: area	92–93, 104–105
17	rectangle: perimeter	88–89, 98–99
18	parallelogram: area	92–93, 108–109
19	circle: circumference	110–111
20	circle: area	112–113
21	two-step area problem	118–121
22	length, area, and volume units	78–79, 96–97, 131–132
23	rectangular solid: volume	133–134
24	cylinder: volume	135

Building
Number
Power

ANGLES

Angles Skills Inventory

This inventory will let you know what you need to work on in the Angles section of *Geometry*. Do all of the following problems that you can. There is no time limit.

Problems 1–3: Write the name of each angle.

1. Symbol: _____

2. Symbol: _____

3. Symbol: _____

Problems 4 and 5: Circle the letter of the correct response.

4. Angle ABC is a(n) _____ angle.

 a. acute
 b. right
 c. obtuse
 d. straight
 e. reflex

5. Angle RST is a(n) _____ angle.

 a. acute
 b. right
 c. obtuse
 d. straight
 e. reflex

6. Measure ∠LMN.

 ∠LMN = _____

7. Measure ∠XYZ.

 ∠XYZ = _____

8. Using the protractor below, draw a 110° angle that opens counter-clockwise ⌢ .

9. Using the protractor below, draw a 75° angle that opens clockwise ⌢.

10. What is the measure of ∠GHI in the drawing below?

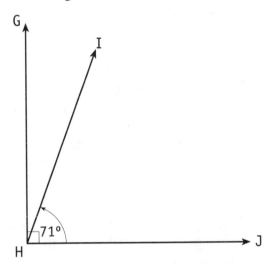

11. In the drawing below, a brace makes a 37° angle with the cross beam. What is the measure of the complementary angle that the brace makes with the upright beam?

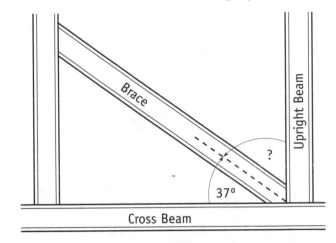

12. What is the complement of 14°?

13. A fireplace hearth is designed to divide the corner angle (90°) of a room into two smaller angles. If the larger of the two complementary angles is 58°, what does the smaller angle measure?

14. What is the measure of ∠ABC in the drawing below?

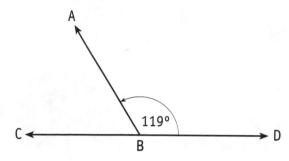

15. Oak Street intersects Twenty-second Avenue at an obtuse angle of 126°. What is the measure of the acute angle at the intersection of Oak and Twenty-second?

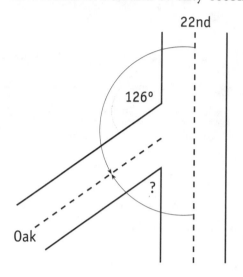

16. What is the supplement of 138°?

17. Alan placed a ladder against the side of his house. The ladder makes an acute angle of 78° with the ground. What is the measure of the supplementary angle that the ladder makes with the ground?

18. What are the measures of ∠2, ∠3, and ∠4 in the drawing below?

∠2 = _____

∠3 = _____

∠4 = _____

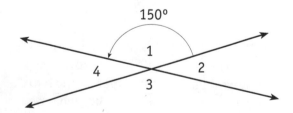

19. Lincoln Street intersects Howard Street at an acute angle of 25°. What are the measures of the other angles at this intersection?

∠a = _____

∠b = _____

∠c = _____

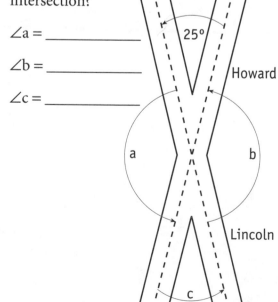

20. In the figure below there are seven unmeasured angles. Label each of these angles with its measure. (A ‖ B)

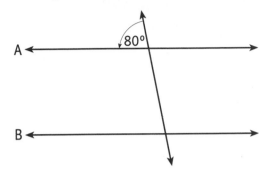

21. The Amtrak railroad tracks cross Elm and Pine Streets as shown in the drawing below. Label each angle at both intersections with its measure.

Angles Inventory Chart

Use this inventory to see what you already know about angles and what you need to work on. A passing score is 18 correct answers. Even if you have a passing score, circle the number of any problem that you miss and turn to the practice pages indicated.

PROBLEM NUMBERS	SKILL AREA	PRACTICE PAGES
1, 2, 3	labeling angles	20
4, 5	types of angles	18–19
6, 7	measuring angles	21–23
8, 9	drawing angles	24–25
10, 12, 14, 16, 18	pairs of angles	26–27
11, 13, 15, 17, 19	pairs of angles: applying skills	28–29
20	parallel lines cut by transversal	30–31
21	parallel lines: applying skills	32–33

INTRODUCING ANGLES

A **line** is a path of points that extend in opposite directions.

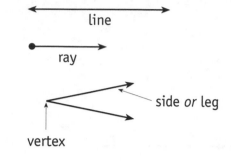

A **ray** is part of a line. A ray has one endpoint.

An **angle** is formed when two rays meet at a point.

- The rays that form an angle are called the **sides** or **legs** of the angle.
- The point where the rays meet is called the **vertex**.

The **size** of an angle depends only on the opening between its sides.

Angles can be very small or very large.

Notice that an arc ⌒ is drawn to tell which angle is indicated.

For example,

might represent either

or

Notice also that the size of an angle is not changed when an angle is turned or when the lengths of the sides are changed.

Each angle below is the same size.

Turning an angle does not change its size.

Changing the length of one or both sides of an angles does not change its size.

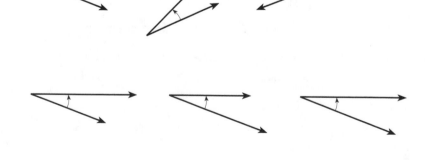

Circle the largest angle in each group of three angles below.

1.

2.

3.

4.

5.

6.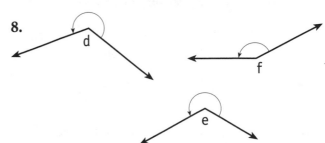

In each group of three angles below, circle the two equal angles.

7.

8.

9.

10.

Measurement of an Angle

As you've seen, the size of an angle depends only on the opening between its sides. This opening is measured in units called **degrees.** The symbol of degrees is °. Thus, 45 degrees is written 45°. A larger angle has a wider opening with more degrees than a smaller angle.

Degrees can be thought of as parts of a circle. A whole circle is divided into 360 degrees. Thus, one degree is $\frac{1}{360}$ of a circle.

—————————— 1°

The size of an angle is found by comparing the angle with a divided circle:

One side of the angle can be placed along the 0° line. The point where the second side crosses the circle shows the number of degrees in the angle.

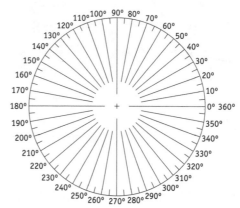

A circle contains 360°.

EXAMPLE 1 What is the size of the angle at the right?

One side of the angle is placed along the 0° line. The second side crosses the circle at **80°.**

ANSWER: The angle measures 80°.

EXAMPLE 2 A pie is cut into eight equal pieces. What angle do the straight sides of each piece make?

A whole pie is a circle. Since the pie is cut into eight equal pieces, each piece contains $\frac{1}{8}$ of the total degrees in a circle.
To find the number of degrees in each piece, divide 360° by 8.
Each piece = $\frac{360°}{8}$ = **45°**

ANSWER: Each piece is a 45° angle.

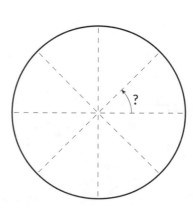

How many degrees are in each angle below?

1. _____

2. _____

3. _____

Match each angle with its measure (size). For reference, use the divided circle on page 16.

4. _____

a. 180°

5. _____

b. 150°

6. _____

c. 210°

7. _____

d. 35°

8. _____

e. 90°

9. A whole pie is divided into twelve equal pieces. At what angle is each piece cut?

10. A circular loaf of bread is to be divided into nine equal pieces. At what angle should each piece be cut?

11. How many degrees are in one-fourth of a circle?

12. How many degrees are in an angle that divides a circle into two equal parts?

Types of Angles

An angle is often named according to its size. Below are listed five types of angles with which you should become familiar.

ACUTE ANGLES

An **acute angle** measures more than 0° but less than 90°.

EXAMPLES

Less than 90°

Acute Angles

More than 0°

RIGHT ANGLES

A **right angle** measures exactly 90°. A right angle is often indicated by placing a small square at its vertex.

EXAMPLES

Exactly 90°

90°

Right Angle

0°

> **Note:** Lines that meet at a right angle are called perpendicular lines. The symbol for perpendicular lines is ⊥.

OBTUSE ANGLES

An **obtuse angle** measures more than 90° but less than 180°.

EXAMPLES

More than 90°

Obtuse Angles

Less than 180°

STRAIGHT ANGLES

A **straight angle** measures exactly 180°. A straight angle is the same as a straight line.

Exactly 180°

Straight Angle

EXAMPLES

REFLEX ANGLES

A **reflex angle** measures more than 180° but less than 360°.

EXAMPLES

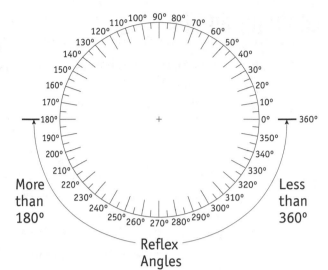

More than 180°

Less than 360°

Reflex Angles

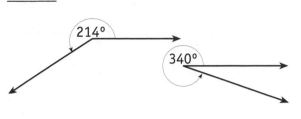

..

Note: A reflex angle is the outside angle formed when two lines meet at a point.

..

Name each angle below: acute, right, obtuse, straight, or reflex.

1. _____

28°

2. _____

3. _____

150°

4. _____

180°

5. _____

310°

6. _____

88°

7. _____

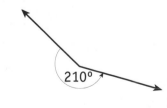
131°

8. _____

210°

9. _____

Labeling an Angle

Angles are often labeled with letters or numbers. The symbol ∠ is commonly used to stand for the word *angle*.

Two labeling methods are shown below.

Labeling Methods **Symbol**

1. An angle is often labeled with three letters: one letter is placed at the end of each side, and one is placed at the vertex.
 The letter placed at the vertex is called the **vertex letter.** The vertex letter is always written as the second of the three.
 Note: The angle at right can be written as ∠ABC or as ∠CBA.

 1. By three letters

 ∠ABC
 or
 ∠CBA

2. An angle can also be labeled with a vertex letter or number only. The vertex letter or number can be placed either on the outside or the inside of the angle.

 2. By vertex letter or number

Outside of angle: B ∠B

Inside of angle: 3 ∠3

Write the symbol that represents each labeled angle below.

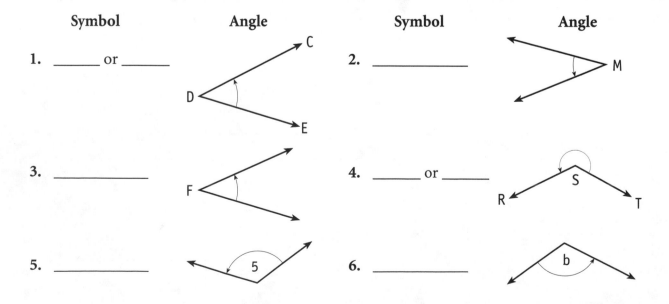

Symbol	Angle	Symbol	Angle
1. _____ or _____		2. _____	
3. _____		4. _____ or _____	
5. _____		6. _____	

Measuring an Angle with a Protractor

A **protractor** is a tool used for measuring and drawing angles. The most commonly used protractor is shaped like a half circle and can measure angles up to 180°. Carpenters, machinists, and others who read blueprints or plans use protractors in their work.

EXAMPLE Use a protractor to measure angle RST at the right.

PROTRACTOR

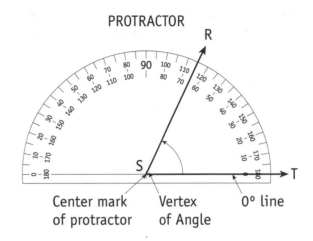

STEP 1 Place the center mark of the protractor at the vertex of the angle, and place the 0° line along one side (ST) of angle RST.

STEP 2 Read the point where the second side (SR) crosses the protractor scale.

ANSWER: ∠RST = 65°

Measure and name the type of each angle drawn below.

1. ∠ABC = _____
 Type of Angle: _____

2. ∠DEF = _____
 Type of Angle: _____

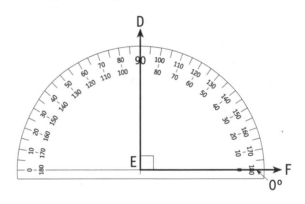

3. ∠XYZ = _____
 Type of Angle: _____

4. ∠RST = _____
 Type of Angle: _____

You can identify an angle by an arc that points either clockwise ⌢ or counterclockwise ⌢. So far in the study of angles, you have identified each angle as opening counterclockwise. However, many angles are easier to measure if you identify them as opening in a clockwise direction.

To make measuring easier, a protractor usually has two scales. The inside scale is used to measure angles that open counterclockwise, and the outside scale is used to measure angles that open clockwise. Be careful to read the correct scale for the angle you are measuring.

EXAMPLE 1 Angle LMN is an acute angle that opens counterclockwise.
∠LMN measures 65°

Use this scale for angles that open counterclockwise.

EXAMPLE 2 Angle ABC is an acute angle that opens clockwise.
∠ABC measures 40°

Use this scale for angles that open clockwise.

Measure each angle and circle the letter of the correct response.

5. ∠ABC measures _____.

The name given to ∠ABC is

a. acute
b. right
c. obtuse
d. straight
e. reflex

6. ∠RST measures _____.

The name given to ∠RST is

a. acute
b. right
c. obtuse
d. straight
e. reflex

7. ∠DEF measures _____.

The name given to ∠DEF is

 a. acute
 b. right
 c. obtuse
 d. straight
 e. reflex

8. ∠LMN measures _____.

The name given to ∠LMN is

 a. acute
 b. right
 c. obtuse
 d. straight
 e. reflex

9. ∠XYZ measures _____.

The name given to ∠XYZ is

 a. acute
 b. right
 c. obtuse
 d. straight
 e. reflex

10. ∠CDE measures _____.

The name given to ∠CDE is

 a. acute
 b. right
 c. obtuse
 d. straight
 e. reflex

Drawing an Angle with a Protractor

To draw an angle with a protractor, draw one side along the 0° line. Draw the other side crossing the protractor scale at the desired mark.

EXAMPLE Using a protractor, draw a 20° angle that opens counterclockwise ⌢.

STEP 1 Draw one side of the angle along the 0° line to the right of the center mark of the protractor.

STEP 2 Draw the second side to cross the protractor scale at 20° on the counterclockwise scale, the inner scale.

Counterclockwise Scale

Draw each angle on the protractor and answer each question.

1. Draw an angle of 75° that opens counterclockwise ⌢.

 If the sides of a 75° angle are doubled in length, the size of the angle

 a. doubles to 150°
 b. is halved to $37\frac{1}{2}°$
 c. stays the same
 d. is increased to 300°
 e. none of the above

2. Draw a 162° angle that opens clockwise ⌢.

 A 162° angle is _____ than a straight angle.

 a. 62° smaller
 b. 18° smaller
 c. 90° larger
 d. 28° larger
 e. none of the above

3. Draw an angle of 40° that opens clockwise ⌒.

 A 40° angle is equal to _____ of a complete circle.

 a. $\frac{1}{2}$

 b. $\frac{1}{6}$

 c. $\frac{1}{9}$

 d. $\frac{1}{3}$

 e. $\frac{1}{10}$

4. Draw a 125° angle that opens counterclockwise ⌒.

 An angle measuring 125° is _____ than a right angle.

 a. 35° smaller
 b. 25° larger
 c. 55° larger
 d. 35° larger
 e. 25° smaller

5. Draw a right angle that opens counterclockwise ⌒.

 An angle that is half the size of a right angle measures

 a. 90°
 b. 30°
 c. 45°
 d. 180°
 e. 145°

Pairs of Angles

As you've seen, a special name is given to an angle according to its size. Also, names are given to special pairs of angles. The names and relationships you should become familiar with are **complementary angles, supplementary angles,** and **vertical angles.**

COMPLEMENTARY ANGLES

Complementary angles are angles whose sum is 90°.

In the drawing at the right, line TR divides right angle QRS into the two complementary angles, $\angle QRT$ and $\angle TRS$.

$\angle QRT$ is called the **complement** of $\angle TRS$.

Often, the value of one complementary angle is given, and you are asked to find the value of the other.

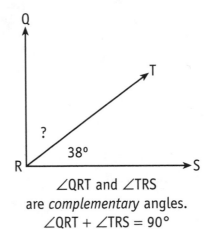

$\angle QRT$ and $\angle TRS$
are *complementary* angles.
$\angle QRT + \angle TRS = 90°$

EXAMPLE Find $\angle QRT$ if $\angle TRS = 38°$.

To find the value of a complementary angle, subtract the known angle from 90°.

$$\angle QRT = 90° - 38° = 52°$$

ANSWER: $\angle QRT = 52°$

 If you use a calculator, $\boxed{C}\;\boxed{9}\;\boxed{0}\;\boxed{-}\;\boxed{3}\;\boxed{8}\;\boxed{=}\;\boxed{\qquad 52.}$

SUPPLEMENTARY ANGLES

Supplementary angles are angles whose sum is 180°.

In the drawing at right, line RC divides the straight angle BCD into the two supplementary angles $\angle BCR$ and $\angle RCD$.

$\angle BCR$ is called the **supplement** of $\angle RCD$.

EXAMPLE Find $\angle BCR$ if $\angle RCD = 130°$.

To find the value of a supplementary angle, subtract the value of the known angle from 180°.

$$\angle BCR = 180° - 130° = 50°$$

ANSWER: $\angle BCR = 50°$

$\angle BCR$ and $\angle RCD$
are *supplementary* angles.
$\angle BCR + \angle RCD = 180°$

VERTICAL ANGLES

Vertical angles are formed when two straight lines cross at a point.

Vertical angles lie across from each other and are equal. In the drawing at right, ∠1 and ∠3 are vertical angles, and ∠2 and ∠4 are vertical angles.

Vertical angles are equal.
∠1 = ∠3 and ∠2 = ∠4

EXAMPLE If ∠1 = 50°, then ∠3 = 50°, and if
∠2 = 130°, then ∠4 = 130°.

Note: When two lines cross, each acute angle is supplementary to each obtuse angle. For example, ∠1 + ∠2 = 180°.

Find the measure of each angle indicated below.

1. ∠ABD = _____

2. ∠TRS = _____

3. ∠XYO = _____

4. ∠MNP = _____

5. ∠DBC = _____

6. ∠OPT = _____

7. ∠b = _____
∠c = _____
∠d = _____

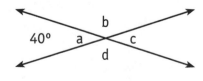

8. ∠1 = _____
∠3 = _____
∠4 = _____

9. ∠a = _____
∠b = _____
∠c = _____

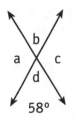

Pairs of Angles: Applying Your Skills

Practical problems involving pairs of angles often require you to find an unknown angle. Examples that frequently appear include

- finding an angle between a ladder and the ground

- finding an angle between two boards or beams

- finding an angle between intersecting streets

- finding a missing corner-of-the-room angle

In each case, you are really being asked to find the complement or supplement of a given angle or to recognize an equal vertical angle.

In books and on tests, practical problems often appear as word problems. You may or may not be given an accompanying drawing. In either case, the first step in solving a word problem is to represent the information in the problem by a drawing that contains only straight lines. Then, solve the straight-line drawing by using the skills you learned on the previous few pages.

EXAMPLE A ladder leans against the side of a house. The ladder makes an acute angle of 61° with the ground. What is the measure of the obtuse angle the ladder makes with the ground?

In this word problem, no drawing is given.
For comparison, look at both an accurate picture and a straight-line drawing.

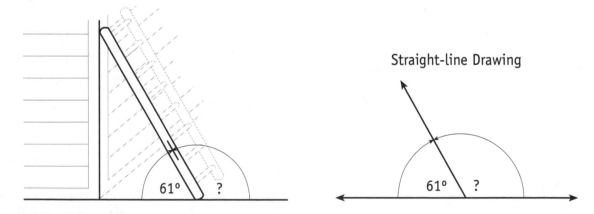

Notice that the use of a straight-line drawing simplifies the problem by focusing attention on the unknown angle. The question really asks,
 "What is the supplement of 61°?"
Find the supplement of 61° by subtracting 61° from 180°.
 Supplement of 61° = 180° − 61° = **119°**

ANSWER: The obtuse angle is 119°.

For each drawing below, solve for the unmeasured angle.

1. The diagonal bridge support piece makes an angle of 50° with the cross beam. What angle does it make with the upright beam?

2. Ohio Street intersects Seventh Street as shown below. What are the measures of the three unmeasured angles?

∠1 = _____

∠2 = _____

∠3 = _____

In problems 3 through 5, make a line drawing if you find it helpful.

3. A ladder is leaning against the side of a barn. If the ladder makes an acute angle of 82° with the ground, what is the measure of the supplementary angle the ladder makes with the ground?

4. Four angles are formed when two straight lines cross. If one of the angles is 90°, what is the measure of each of the other angles?

 5. Estimate the supplement of 148.5°.

Refer to the drawing at the right, and match each expression with its equivalent measure.

Expression		Measure
_____ **6.** ∠1		**a.** 180°
_____ **7.** ∠4 + ∠5		**b.** 90°
_____ **8.** ∠5		**c.** 360°
_____ **9.** ∠1 + ∠2		**d.** 45°
_____ **10.** ∠1 + ∠2 + ∠3 + ∠4 + ∠5		**e.** 135°

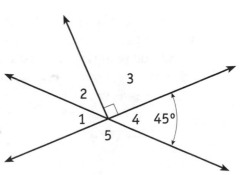

Parallel Lines Cut by a Transversal

Parallel lines are lines that run side-by-side and do not cross at any point. Lines A and B in the drawing below are parallel lines.

A **transversal** is a third line that crosses two parallel lines. The line labeled *t* in the drawing is a transversal.

When a transversal crosses two parallel lines, a set of four angles is formed at each point of intersection.

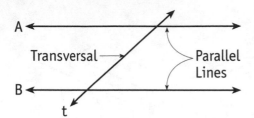

Note: The symbol for parallel is ∥. To write "Line A is parallel to Line B", write A ∥ B.

Since one set of angles looks exactly like the other set of angles, the following relations are true:

I. *Corresponding Angles Are Equal*
 If you cut the transversal in half and place one set of angles over the other, corresponding angles would sit on top of each other.

 $\angle 1 = \angle 5$ \qquad $\angle 3 = \angle 7$
 \qquad and
 $\angle 2 = \angle 6$ \qquad $\angle 4 = \angle 8$

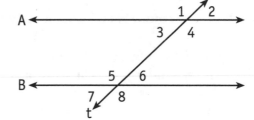

II. *Alternate Interior Angles Are Equal*
 Alternate interior angles are angles that are inside (interior) the parallel lines and that are on opposite (alternate) sides of the transversal.

 $\angle 3 = \angle 6$
 $\angle 4 = \angle 5$

III. *Alternate Exterior Angles Are Equal*
 Alternate exterior angles are angles that are outside (exterior) the parallel lines and that are on opposite (alternate) sides of the transversal.

 $\angle 1 = \angle 8$
 $\angle 2 = \angle 7$

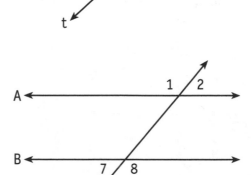

You will seldom need to remember the definitions introduced on the previous page. However, you should remember the angle relationships when two parallel lines are cut by a transversal. These relationships are most easily remembered as follows.

- The four acute angles are equal.

- The four obtuse angles are equal.

- Each acute angle is supplementary to each obtuse angle.

<u>**EXAMPLE**</u> Parallel lines E and F are cut by transversal *r*. The acute angle at the upper right is 50°. Label the other seven angles in the figure.

STEP 1 Label the acute angle at the upper right as 50°.

STEP 2 Label all other acute angles as 50°.

STEP 3 The supplement of 50° is found by subtracting 50° from 180°.
$$180° - 50° = 130°$$
Label each obtuse angle as 130°.

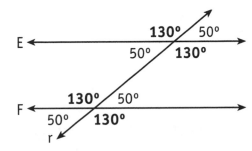

In the figures below, label each angle with its measure.

1. A ∥ B

2. M ∥ N

3. A ∥ B

(**Hint:** Extend the transversal across each parallel line.)

4. C ∥ D

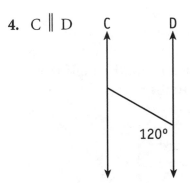

(**Hint:** Don't be fooled just because the figure is turned.)

Parallel Lines Cut by a Transversal: Applying Your Skills

In each drawing below, solve for the unmeasured angle as indicated.

1. Jackson Street crosses Twenty-third and Twenty-fourth Streets at an angle of 70°. What is the measure of angle c in the drawing below?

2. The diagonal brace in the fence gate makes an angle of 50° with the vertical side piece on the left. What angle does the brace make with the vertical side piece on the right?

3. The commuter train crosses Monroe and Madison Avenues as shown below. What is the measure of the angle labeled *a* in the drawing?

4. Lincoln Street joins Eighth and Ninth as shown below. How many degrees are in angle D, the obtuse angle that Lincoln makes with Ninth?

5. The side of the A-frame house below makes an angle of 130° with the ground. What is the measure of the angle that the side makes with the rafter?

6. A parallelogram is a four-sided figure with two pairs of parallel lines. Find the sum of the four angles of a parallelogram. (**Hint:** Each obtuse angle is supplementary to each acute angle.)

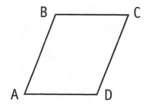

7. The Big H Clothing Company designed a new symbol to place on its products. The designer suggested that all acute angles should be 60°. What is the measure of each obtuse angle?

BIG H
Clothing Company

8. The capital letter I is an example of parallel lines joined by a transversal. Name four other capital letters that are similar examples.

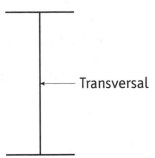

Transversal

Parallel lines often occur in patterns. Don't let these patterns confuse you. Look for the parallel lines and transversals in the drawings below. Notice that you are asked to find an angle that is equal to or the supplement of a given angle.

9. In the drawing below showing two sets of three parallel lines, which angle (∠b or ∠c) is equal to ∠a?

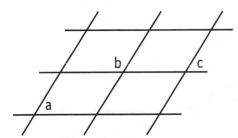

10. A crochet pattern on the edge of a baby blanket is designed as a zigzag between parallel lines. In the drawing below, what is the measure of angle w? ∠v = 60°

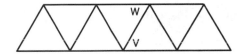

11. The shelf brace below makes an obtuse angle of 145° with the lower shelf. What acute angle does the brace make with the upper shelf?

12. Cross braces are often used in the framing of a house. In the drawing below, estimate the measure of the angle labeled b.

ANGLES REVIEW

···

This review covers the material you have just studied.

Problems 1–3: Write the name of each angle.

1. Symbol: _____

2. Symbol: _____

3. Symbol: _____

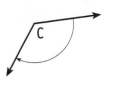

Problems 4 and 5: Circle the letter of the correct response.

4. Angle CDE is a(n) _____ angle.

 a. acute
 b. right
 c. obtuse
 d. straight
 e. reflex

5. Angle XYZ is a(n) _____ angle.

 a. acute
 b. right
 c. obtuse
 d. straight
 e. reflex

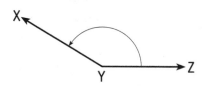

6. Measure ∠FGH.

 ∠FGH = _____

7. Measure ∠QRS.

 ∠QRS = _____

8. Use the protractor below and draw a 165°
angle that opens counterclockwise ⌒.

9. Use the protractor below and draw a 45°
angle that opens clockwise ⌒.

10. What is the measure of ∠CDE in the
drawing below?

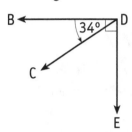

11. As shown in the drawing below, a mirror
frame has a 60° angle in the corner. What
is the measure of the complementary
angle a in the drawing?

12. What is the complement of 71°?

13. A door, placed next to a corner of a room,
is open at an 80° angle, almost touching a
wall. What is the measure of the
complementary angle that the door
makes with the wall?

14. What is the measure of ∠TUV in the
drawing below?

15. The shelf support, drawn below, forms a
40° acute angle with the wall. How large is
the obtuse angle that the support forms
with the wall?

16. What is the supplement of 143°?

17. Webster Street intersects Thirty-fourth Street at a 65° acute angle. What is the measure of the supplementary angle that Webster makes with Thirty-fourth?

18. What are the measures of ∠a, ∠b, and ∠c in the drawing below?

∠a = _____

∠b = _____

∠c = _____

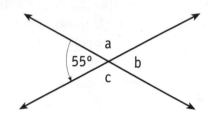

19. The two walking paths in Central Park cross near the fountain as shown in the drawing below. What are the measures of the other three angles at this intersection?

∠1 = _____

∠2 = _____

∠3 = _____

20. In the figure below, label each angle with its measure. (A ∥ B)

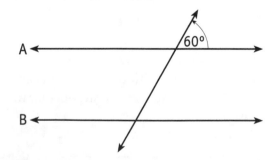

21. Front Street intersects Fifty-first at an angle of 62° as shown. Label each angle at both intersections with its measure.

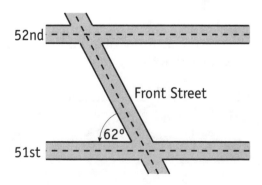

Angles Review Chart

Circle the number of any problem that you missed and review the appropriate pages. A passing score is 18 correct answers. If you miss more than three questions, you should review this chapter.

PROBLEM NUMBERS	SKILL AREA	PRACTICE PAGES
1, 2, 3	labeling angles	20
4, 5	types of angles	18–19
6, 7	measuring angles	21–23
8, 9	drawing angles	24–25
10, 12, 14, 16, 18	pairs of angles	26–27
11, 13, 15, 17, 19	pairs of angles: applying skills	28–29
20	parallel lines cut by transversal	30–31
21	parallel lines: applying skills	32–33

TRIANGLES

Triangles Skills Inventory

This inventory will let you know what you need to work on in the Triangles section of *Geometry*. Do all of the following problems that you can. There is no time limit.

1. What is the measure of ∠B in the triangle below?

2. What is the top roof angle (angle a) in the drawing below?

3. The sum of two angles in a triangle is 117°. What is the measure of the third angle?

In problems 4 and 5, circle the letter of the correct response.

4. The name given to triangle DEF is

 a. equilateral
 b. isosceles
 c. right
 d. scalene
 e. isosceles and scalene

5. The name given to triangle QRS is

 a. equilateral
 b. isosceles
 c. right
 d. scalene
 e. right and scalene

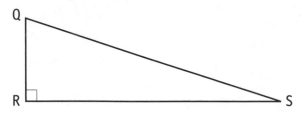

6. Triangles ABC and XYZ are similar triangles. What is the length of side YZ in triangle XYZ?

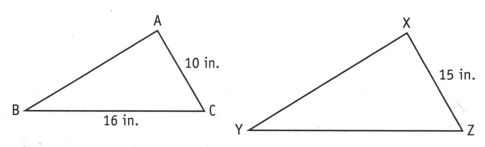

7. In triangle OPQ, ∠O = 42° and ∠P = 31°.
 In triangle XYZ, ∠X = 31° and ∠Y = 107°.
 Is triangle OPQ similar to triangle XYZ?

8. To find the distance (*d*) across Boot Lake, Anna measured the distances as shown on her drawing. Note that the two triangles are similar, and find the distance for *d*. (Distances are measured in yards.)

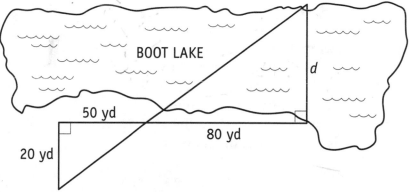

9. To find the height of a fir tree, Bill first found the length of the tree's shadow to be 80 feet long. At the same time, he held a yardstick perpendicular to the ground. The yardstick cast a shadow of 4 feet. What was the height of the tree? (**Hint:** How many feet are in a yardstick?)

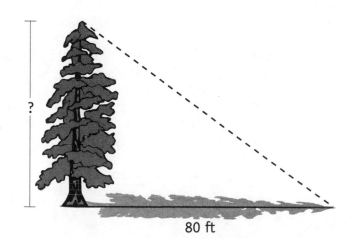

10. Find the square of each number.

 a. $3^2 =$ _____

 c. $7^2 =$ _____

 b. $5^2 =$ _____

 d. $9^2 =$ _____

11. Find each square root.

 a. $\sqrt{4} =$ _____

 c. $\sqrt{64} =$ _____

 b. $\sqrt{36} =$ _____

 d. $\sqrt{100} =$ _____

12. What is the length of the hypotenuse of $\triangle ABC$?

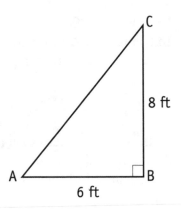

13. As shown in the drawing, the support piece for the billboard attaches to the billboard 12 feet above the ground. The support piece rests on the ground 5 feet from the base of the billboard. What is the length of the support piece?

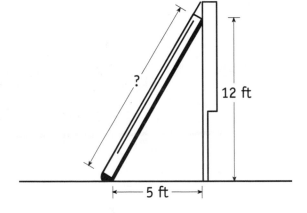

14. One side of a right triangle measures 15 feet. The hypotenuse measures 17 feet. What is the length of the third side?

Triangles Inventory Chart

Use this inventory to see what you already know about triangles and what you need to work on. A passing score is 13 correct answers. Even if you have a passing score, circle the number of any problem that you miss and turn to the practice pages indicated.

PROBLEM NUMBERS	SKILL AREA	PRACTICE PAGES
1, 3	sum of angles	44–45
2	sum of angles: applying skills	46–47
4, 5	definitions of triangles	48–50
6, 7	similar triangles	52–59
8, 9	similar triangles: applying skills	60–61
10, 11	squares and square roots	62–63
12, 14	Pythagorean theorem	64–67
13	Pythagorean theorem: applying skills	68–69

Introducing Triangles

A **triangle** is a plane (flat) figure that has three sides and three angles. Triangles appear in a variety of shapes and sizes.

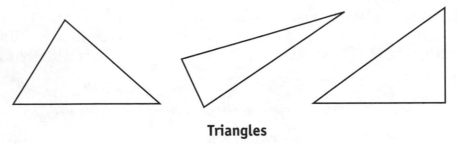

Triangles

LABELING A TRIANGLE

The most common way to label a triangle is with three letters, usually written in alphabetical order. One letter is placed at the vertex of each angle.

The symbol for triangle is Δ—a small triangle with three equal sides.

<u>EXAMPLE</u> The triangle at the right is represented as ΔABC.

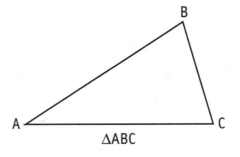

ΔABC

REPRESENTING A SIDE OF A TRIANGLE

The side of a triangle is represented by the two letters at the ends of that side.

<u>EXAMPLE</u> Side ST is the side *opposite* (across from) ∠R.
In ΔRST, side RS is the longest side.

ΔRST

> **Note:** The longest side of a triangle is always opposite the largest angle.

REPRESENTING AN ANGLE IN A TRIANGLE

An angle in a triangle is most often represented by the angle's vertex letter only.

<u>EXAMPLE</u> In ΔDEF, the 30° angle is labeled D and usually identified as ∠D.

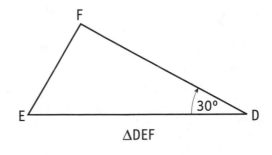

ΔDEF

> **Note:** Angle D can also be called ∠EDF or ∠FDE. When three letters are used, the vertex letter is written second.

Write the letters (in alphabetical order) that represent each triangle.

1. _____

2. _____

3. _____

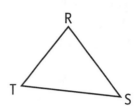

Name the side that is opposite each measured angle.

4. _____

5. _____

6. _____

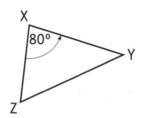

Using a single letter, name the angle that is opposite a measured side.

7. _____

8. _____

9. _____

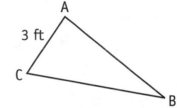

Referring to the triangle below, complete each sentence.

10. The angle opposite the side that has a length of 8 feet is labeled _____.

11. The length of the side opposite ∠a is _____.

12. The angle opposite the side that has a length of 15 feet is labeled _____.

13. The length of the side opposite ∠c is _____.

14. The angle opposite the side that has a length of 12 feet is labeled _____.

15. The largest angle shown is _____.

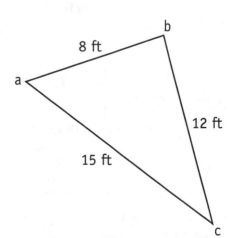

The Sum of Angles in a Triangle

One of the most important rules to learn in the study of geometry is the following:

The sum of the three angles in a triangle is equal to 180°.

You can write this rule in symbols.

$$\angle a + \angle b + \angle c = 180°$$

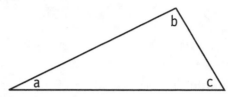

To see why this rule is true, enclose a triangle in parallel lines and use information that you learned in the chapter on angles.

You know that $\angle d + \angle b + \angle e = 180°$ since together they make up a straight angle.

Also, you know that $\angle a = \angle d$ and $\angle c = \angle e$ since these are pairs of opposite interior angles.

Thus, the sum $\angle a + \angle b + \angle c$ is equal to $\angle d + \angle b + \angle e$, which is 180°.

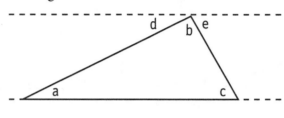

$$\angle a + \angle b + \angle c = 180°$$

The rule stated above is commonly used to find the measure of the third angle in a triangle when the other two are known. To find the measure of the third angle, subtract the sum of the two known angles from 180°.

EXAMPLE 1 What is the measure of $\angle B$ in triangle ABC?

> **STEP 1** Add $\angle A$ and $\angle C$.
>
> $$\angle A + \angle C = 32° + 47° = 79°$$
>
> **STEP 2** To find $\angle B$, subtract 79° from 180°.
>
> $$\angle B = 180° - 79° = \mathbf{101°}$$

ANSWER: $\angle B = 101°$

 If you use a calculator, [C] [1] [8] [0] [−] [3] [2] [−] [4] [7] [=] [101.]

EXAMPLE 2 What is the measure of $\angle D$ in the right triangle DEF?

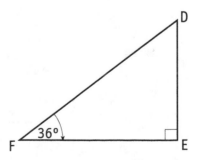

> **STEP 1** Add $\angle E$ and $\angle F$. Remember: $\angle E = 90°$.
>
> $$\angle E + \angle F = 90° + 36° = 126°$$
>
> **STEP 2** To find $\angle D$, subtract 126° from 180°.
>
> $$\angle D = 180° - 126° = 54°$$

ANSWER: $\angle D = 54°$

Find the measure of the third angle in each triangle.

1. _____

2. _____

3. _____

4. _____

5. _____

6. _____

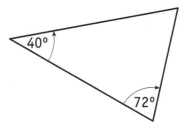

7. _____

8. _____

9. _____

10. In △RST, ∠R = 45° and ∠S = 52°. What is the measure of ∠T?

11. Angle C in △ABC is a right angle. If ∠B is equal to 35°, what is the measure of ∠A?

12. In △XYZ, each of two angles is equal to 39.75°. Estimate the measure of the third angle.

The Sum of Angles in a Triangle: Applying Your Skills

Knowing that a triangle contains 180° is useful in finding the third angle of a triangle when the first two are known.

EXAMPLE The support cable for a telephone pole makes an angle of 63° with the ground. What angle does the cable make with the pole?

> **Note:** The telephone pole is assumed to make a 90° angle with the ground.

STEP 1 Add 63° and 90°.

$$63° + 90° = 153°$$

STEP 2 To find the unknown angle, subtract 153° from 180°.

$$180° - 153° = 27°$$

ANSWER: **The cable makes an angle of 27° with the pole.**

If you use a calculator, C 1 8 0 − 6 3 − 9 0 = ⟦ 27.⟧

1. A shelf support makes an angle of 48° with the wall. What is the measure of the angle the support makes with the shelf?

2. To properly cut her greenhouse roof side piece, what is the measure of the third angle Joyce needs to know?

3. Carey leaned a ladder against the house. If the ladder makes an angle of 62° with the ground, what is the measure of the angle it makes with the house?

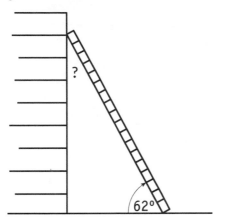

4. What is the measure of the roof angle in the drawing below?

5. The walking paths in City Park meet in a triangle around the rose garden. What is the measure of the third angle in this triangle?

6. A dress design calls for a triangular piece of material. If one angle is a right angle and a second angle is 35°, what is the measure of the third angle?

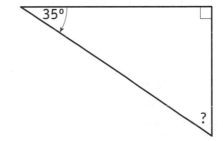

7. A quilt pattern requires pieces of cloth cut in the shape of a triangle. Each triangle is to have three equal angles and three equal sides. What is the measure of each of the three equal angles in a triangle?

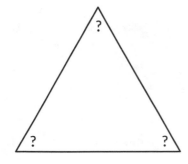

8. A surveyor drove a stick into the ground. He then dropped a line from the top of the stick and anchored it into the ground in front of the stick. If the angles shown below are correct, does the stick make a 90° angle with the ground?

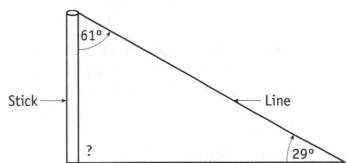

Working with Definitions of Triangles

You will read about three types of triangles: equilateral, isosceles, and scalene. Knowing the characteristics of these types of triangles is necessary in solving many types of problems. In this section, definitions and examples of problems you should be familiar with are given.

EQUILATERAL TRIANGLE

An **equilateral triangle** has three equal angles. Each angle measures 60°, and all three sides are equal.

An equilateral triangle is also called an **equiangular triangle.**

<u>EXAMPLE</u> What is the measure of each angle in the triangle at the right?

Since all three sides are equal, the triangle is equilateral. Thus, all three angles are equal, and each angle has a measure of 60°.

$$180° \div 3 = \mathbf{60°}$$

ANSWER: Each angle is 60°.

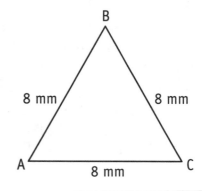

Equilateral Triangle

ISOSCELES TRIANGLE

An **isosceles triangle** has two equal angles. The sides opposite the equal angles are also equal.

The two equal angles are called **base angles.** The third angle is called the **vertex angle.**

<u>EXAMPLE 1</u> What is the value of each base angle in the isosceles triangle at the right?

STEP 1 The sum of the base angles is found by subtracting 50° from 180°.

$$180° - \angle y = \angle x + \angle z$$
$$180° - 50° = 130° \ (\angle x + \angle z)$$

STEP 2 Since $\angle x$ and $\angle z$ are equal, divide your answer from the first step by 2. This will give the value of each base angle.

$$130° \div 2 = 65°$$

ANSWER: Each base angle is 65°.

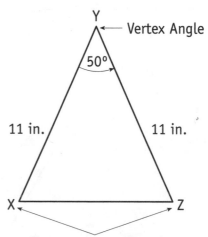

Base angles are equal.

Isosceles Triangle

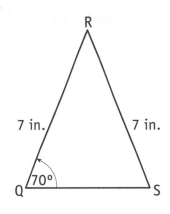

EXAMPLE 2 What is the measure of ∠S and ∠R in the isosceles triangle at the right?

STEP 1 Since this triangle has two equal sides, it must have two equal angles. Thus, the base angles ∠Q and ∠S are equal.

$$∠S = ∠Q = 70°$$

STEP 2 Angle R, the vertex angle, is found by subtracting the sum of the base angles from 180°.

$$∠Q + ∠S = 70° + 70° = 140°$$
$$∠R = 180° - 140° = 40°$$

ANSWER: ∠S = 70° and ∠R = 40°

SCALENE TRIANGLE

A **scalene triangle** has no equal sides and no equal angles. The scalene is the most common type of triangle.

To find a missing angle in a scalene triangle, subtract the two known angles from 180°.

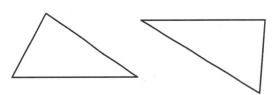

Scalene Triangles

Note: A triangle that has a right angle is often called a **right triangle.** Either an isosceles triangle or a scalene triangle can be a right triangle.

In an isosceles right triangle, one angle is 90°, and the other two angles are equal.

In a scalene right triangle, one angle is 90°, and the other two angles are not equal.

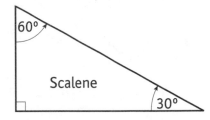

Right Triangles

EXAMPLE What is the value of each base angle in the isosceles right triangle BCD?

STEP 1 The sum of base angles is found by subtracting 90° (∠B) from 180°.

$$180° - ∠B = ∠C + ∠D$$
$$180° - 90° = 90°$$

STEP 2 To find ∠C and ∠D, divide 90° by 2.

$$∠C = ∠D = \frac{90°}{2} = 45°$$

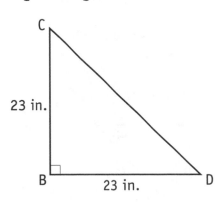

ANSWER: ∠C = ∠D = 45°

In each triangle below:
 a. Write the correct measure of each angle.
 b. Name each triangle as equilateral, isosceles, or scalene.
 c. Circle any triangle that is also a right triangle.

<u>**EXAMPLE**</u> **Problem** **Answer**

Isosceles

1. _____ **2.** _____ **3.** _____

4. _____ **5.** _____ **6.** _____

7. _____ **8.** _____ **9.** _____

**In each isosceles triangle below, find the length of the side labeled
with a question mark.**

10. _____ **11.** _____ **12.** _____

Working With Definitions of Triangles: Applying Your Skills

As a first step in solving each word problem below, draw the triangle that is described and label any angle whose value is given.

1. The vertex angle of an isosceles triangle measures 80°. What is the measure of each base angle?

2. In a scalene right triangle, one angle measures 35°. How large is the second acute angle?

3. Two of the angles in a scalene triangle measure 28° and 57°. What does the third angle measure?

4. If the three sides of triangle XYZ are equal, what is the measure of each angle?

5. If a triangular figure is in the shape of an isosceles right triangle, what is the measure of each base angle?

6. A billiards ball rack is shaped like a triangle with three equal sides. At what acute angle does each pair of sides meet? (**Hint:** An acute angle measures more than 0° but less than 90°.)

7. A child's toy block is in the shape of a triangle. One angle measures 32° and another angle measures 17°. What is the measure of the third angle?

8. The roof gable on Charlene's house has two equal sides. One angle measures 110.4°. The other two angles are equal to each other. Estimate the measure of each of the other two angles.

Similar Triangles

Triangles that have the same three angles are called **similar triangles.** Similar triangles have the same shape and differ only in the length of their sides. The symbol ~ means "is similar to."

EXAMPLE 1 Triangle ABC is similar to triangle DEF.
$$\triangle ABC \sim \triangle DEF$$

Equal angles in similar triangles are called **corresponding angles.** In $\triangle ABC$ and $\triangle DEF$, each pair of corresponding angles is easily identified.

Corresponding Angles

Each triangle has a 35° angle: $\angle A = \angle D$.
Each triangle has a 91° angle: $\angle B = \angle E$.
Each triangle has a 54° angle: $\angle C = \angle F$.

Sides opposite corresponding angles are called **corresponding sides.** Each pair of corresponding sides are opposite equal angles.

Corresponding Sides

Sides opposite the 35° angles: sides BC and EF.
Sides opposite the 91° angles: sides AC and DF.
Sides opposite the 54° angles: sides AB and DE.

EXAMPLE 2 In $\triangle XYZ$, $\angle X = 19°$ and $\angle Y = 76°$.
In $\triangle GHI$, $\angle G = 85°$ and $\angle H = 19°$. Is $\triangle XYZ$ similar to $\triangle GHI$?

STEP 1 To find the third angle in $\triangle XYZ$, subtract the two known angles from 180°.

$$\angle X + \angle Y = 19° + 76° = 95°$$
$$\angle Z = 180° - 95° = 85°$$

STEP 2 To find the third angle in $\triangle GHI$, subtract the two known angles from 180°.

$$\angle G + \angle H = 85° + 19° = 104°$$
$$\angle I = 180° - 104° = 76°$$

STEP 3 Compare the angles of the two triangles.

In $\triangle XYZ$, $\angle X = 19°$, $\angle Y = 76°$, and $\angle Z = 85°$.
In $\triangle GHI$, $\angle G = 85°$, $\angle H = 19°$, and $\angle I = 76°$.

ANSWER: Yes, $\triangle XYZ$ is similar to $\triangle GHI$. Each triangle has the same three angles: 19°, 76°, and 85°.

Questions 1 through 3 refer to triangles LMN and XYZ below.

ΔLMN ~ ΔXYZ

1. Is ΔLMN similar to ΔXYZ?

 Why?

2. What are the three pairs of corresponding angles?

 a. ∠_____ = ∠_____ b. ∠_____ = ∠_____ c. ∠_____ = ∠_____

3. What are the three pairs of corresponding sides?

 a. _____ and _____ b. _____ and _____ c. _____ and _____

Answer the following questions.

4. Of the following three triangles, which two are similar?

5. In ΔLMN, ∠M = 24°, and ∠N = 107°.
 In ΔPQR, ∠P = 107°, and ∠Q = 48°.
 Is ΔLMN similar to ΔPQR?

6. In ΔABC, ∠A = 50°, and ∠B = 42°.
 In ΔDEF, ∠E = 88°, and ∠F = 50°.
 Is ΔABC similar to ΔDEF?

Similar triangles are interesting because the lengths of corresponding sides are in **proportion** and can be written as **equivalent fractions.**

In your earlier work with mathematics, you worked with equivalent fractions; for instance $\frac{2}{4} = \frac{1}{2}$. A proportion is an example of equivalent fractions. In fact, the study of similar triangles is a use of equivalent fractions in geometry.

Example 3 below illustrates how to write corresponding sides as a proportion.

EXAMPLE 3 Because △ABC and △DEF have corresponding angles, they are similar triangles.

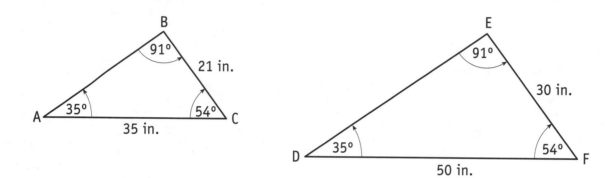

We can write a proportion as follows.

STEP 1 Identify the pairs of corresponding sides.
a. AC and DF are corresponding sides.
b. BC and EF are corresponding sides.

STEP 2 Write equivalent fractions (to be called a proportion from now on).

$$\frac{AC}{DF} = \frac{BC}{EF}$$

$$\frac{35}{50} = \frac{21}{30}$$

(To see that both sides of the proportion are equal, see that both can be reduced to $\frac{7}{10}$.)

In Example 3, it would have also been correct to write the proportion

$$\frac{DF}{AC} = \frac{EF}{BC}$$

$$\frac{50}{35} = \frac{30}{21}$$

In both cases, the corresponding sides are on the same side of the equals sign.

To write the terms of the proportion in correct order, remember
1. Each equivalent fraction consists of corresponding sides.
2. The numerator of each fraction is from one triangle.
3. The denominator of each fraction is from the other triangle.

For each pair of similar triangles below, fill in the missing term for each proportion. The first problem is done as an example.

1. **Problem** **Answer**

$$\frac{12}{36} = \frac{}{27}$$ $$\frac{12}{36} = \frac{9}{27}$$

2. $$\frac{7}{35} = \frac{}{20}$$

3. $$\frac{}{30} = \frac{14}{21}$$

4. $$\frac{15}{12} = \frac{10}{}$$

In problems 5 and 6, don't be fooled because the triangles are turned. Carefully find corresponding sides before completing each proportion.

5. $$\frac{18}{} = \frac{15}{20}$$

6. $$\frac{40}{32} = \frac{}{24}$$

Example 4 shows how similar triangles can be used to find an unknown distance.

EXAMPLE 4 What is length *m* of side LN in ΔLMN below?

STEP 1 Write a proportion using the corresponding sides of triangles RST and LMN.

$$\frac{12}{9} = \frac{8}{m}$$

STEP 2 Solve the proportion by **cross multiplying.** To cross-multiply, follow steps a. and b.

a. Multiply the numerator of each fraction by the denominator of the other fraction. Set the answers equal to each other.

$$\frac{12}{9} \diagdown \diagup \frac{8}{m}$$

$$12m = 72$$

b. Divide the number standing alone by the number next to the *m*. Since 12 times *m* equals 72, find *m* by dividing 72 by 12.

$$m = \frac{72}{12} = 6$$

ANSWER: Side LN has a length of 6 feet.

 Find each length as indicated below.

1. ΔABC ~ ΔRST. What is length *d*?

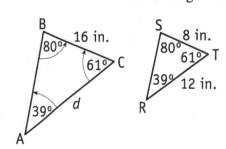

2. ΔDEF ~ ΔKLM. What is length *x*?

 In problems 3 through 8, find corresponding sides before completing each proportion.

3. △ABC ~ △XYZ. What is length *s*?

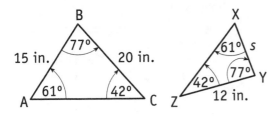

4. △GHI ~ △DEF. What is length *d*?

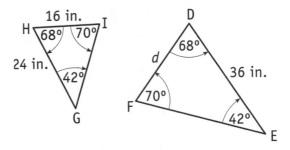

5. △PQR ~ △EFG. What is length *b*?

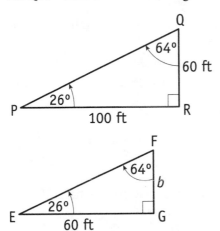

6. △LMN ~ △PQR. What is length *x*?

7. △ABC ~ △DEF. What is length *s*?

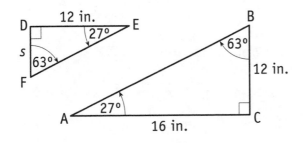

8. △XYZ ~ △GHI. What is length *f*?

There are two geometric figures involving similar triangles that occur often in practical applications and on tests. Study the figures below in Examples 5 and 6 before working problems 1 through 4.

EXAMPLE 5 △RST ~ △RUV

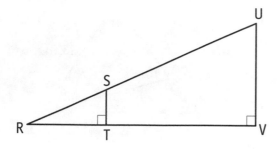

EXAMPLE 6 △ABC ~ △ADE

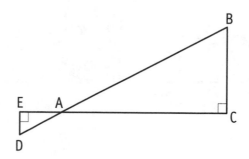

Do you see why △RST is similar to △RUV?

The reason is that each triangle has the same three angles.

a. ∠R is an angle in each triangle.
b. ∠RTS = ∠RVU; each is a right angle.
c. ∠RST = ∠RUV. Since the two other angles are equal in both triangles, the third angle must also be the same.

Sample Question:

If ST = 10 ft, RT = 15 ft, and RV = 60 ft, what is the length of UV?

Solution:

STEP 1 Many students find it helpful to draw the similar triangles separately.

STEP 2 Write a proportion. Let x stand for the unknown value of UV.

$$\frac{UV}{ST} = \frac{RV}{RT} \text{ or } \frac{x}{10} = \frac{60}{15}$$

STEP 3 Solve the proportion for x.

$$15x = 600$$
$$x = \frac{600}{15} = \mathbf{40}$$

ANSWER: UV = 40 ft

Do you see that △ABC is similar to △ADE?

a. ∠EAD = ∠BAC since they are vertical angles.
b. ∠AED = ∠ACB since each is a right angle.
c. ∠D = ∠B. Since the right angles and the vertical angles are equal, the remaining angles must be equal.

Sample Question:

If AC = 100 in., AE = 20 in., and DE = 15 in., what is the length of BC?

Solution:

STEP 1 Again, you may find it helpful to draw the similar triangles separately.

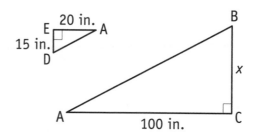

STEP 2 Write a proportion.

$$\frac{AC}{AE} = \frac{BC}{DE} \text{ or } \frac{100}{20} = \frac{x}{15}$$

STEP 3 Solve the proportion for x.

$$20x = 1500$$
$$x = \frac{1500}{20} = \mathbf{75}$$

ANSWER: BC = 75 in.

 Find each length as indicated below.

1. EF = 10 cm, DF = 16 cm, DH = 48 cm
 What is the length of GH?

2. KF = 4 ft, FH = 12 ft, KI = 3 ft.
 What is the length of GH?

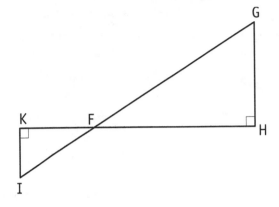

3. LM = 36 mi, MN = 80 mi, PN = 20 mi.
 What is the length of OP?

4. RS = 18 yd, ST = 6 yd, TU = 4 yd.
 What is the length of QR?

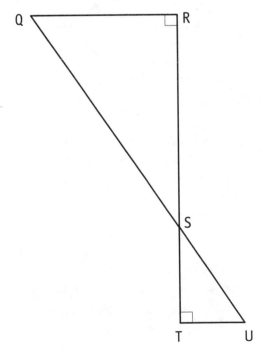

Similar Triangles: Applying Your Skills

Similar triangles are often used to solve practical problems that require you to find an unknown distance or height. The side of a triangle is formed by the height of an object, such as a building, tree, or person, or by the distance across a river or a canyon.

To use similar triangles in a practical problem, follow these basic steps:

STEP 1 Identify the corresponding angles in the similar triangles.

STEP 2 Identify the corresponding sides—opposite the corresponding angles.

STEP 3 Write a proportion involving the unknown distance or height.

STEP 4 Solve the proportion by cross multiplication.

EXAMPLE Two trees stand side-by-side. The smaller tree is 18 feet high and casts a shadow of 30 feet. The larger tree casts a shadow of 55 feet. How high is the larger tree?

55 ft

Note: The sun's rays are parallel. This means that the sides opposite the right angles (indicated by dotted lines in the drawing at right) form equal angles with the ground. $\angle a = \angle c$ and $\angle b = \angle d$.

STEP 1 Since $\angle a = \angle c$, the unknown height h and the height 18 feet are corresponding sides of similar triangles. The shadow lengths are corresponding sides.

$$\frac{h}{18} = \frac{55}{30}$$

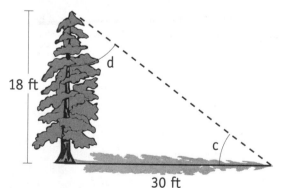

18 ft

30 ft

STEP 2 Solve the proportion by cross multiplying and then dividing.

$$\frac{h}{18} \diagup \!\!\!\!\! \diagdown \frac{55}{30}$$

$$30h = 990$$

$$h = \frac{990}{30} = \mathbf{33}$$

ANSWER: The taller tree is 33 feet high.

Solve each problem for the unknown height or distance.

1. A house stands next to a shed. The shed is 12 feet high and casts a shadow of 9 feet. At the same time, the house casts a shadow of 18 feet. How high is the top of the house?

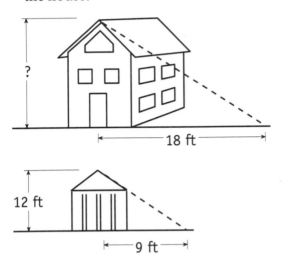

2. To measure the distance (*d*) across Green Lake, Bill marked off distances and drew the picture below. Use the similar triangles to solve for the distance across Green Lake.

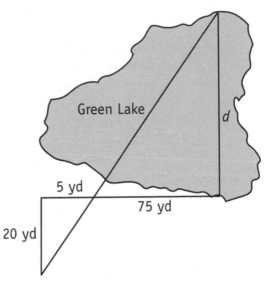

3. Joyce is standing next to a telephone pole. She walks along the shadow of the pole until the top of her shadow meets the top of the shadow of the pole. Joyce is 5 feet tall, and her shadow measures 8 feet. The shadow of the pole measures 112 feet. How tall is the pole?

4. Laurie estimates the distance across Blue River by walking off distances and making the drawing below. Use her drawing to find the distance (*d*) across the river.

Squares and Square Roots

Before beginning the next topic, the Pythagorean theorem, spend some time on these two pages strengthening your skills working with **squares** and **square roots.**

SQUARES

The **square** of a number is that number multiplied by itself. For example, the square of 5 is $5 \times 5 = 25$.

In symbols, you write the square of a number as a **base** and an **exponent.**

5×5 is written 5^2
exponent
base

The **exponent** (2) tells how many times the **base** (5) is written in the product when multiplying. You read 5^2 as five squared. The **value** of 5^2 is 25.

Look at these examples.

Product	As a base and an exponent	Read in words	Value
4×4	4^2	4 squared	16
11×11	11^2	11 squared	121
$b \times b$	b^2	b squared	*
$c \times c$	c^2	c squared	*

*A value for a letter squared can only be found when we know the letter's value.

EXAMPLE What is the value of a^2 if $a = 7$?

Replace the letter a with the number 7.

$a^2 = 7^2 = 7 \times 7 = \mathbf{49}$

ANSWER: $a^2 = 49$ if $a = 7$

What is the value of each squared number or letter below?

1. $3^2 =$ _____

2. $8^2 =$ _____

3. $6^2 =$ _____

4. $9^2 =$ _____

5. $12^2 =$ _____

6. $10^2 =$ _____

7. $14^2 =$ _____

8. $15^2 =$ _____

9. a^2 if $a = 3$ _____

10. b^2 if $b = 7$ _____

11. c^2 if $c = 4$ _____

12. d^2 if $d = 17$ _____

13. x^2 if $x = 18$ _____

14. b^2 if $b = 13$ _____

SQUARE ROOTS

The second skill to learn is finding a **square root.**

The square root of a number is found by asking, "What number times itself equals this?" For instance, what number times itself equals 4? The answer is 2 because $2 \times 2 = 4$. This is called finding the square root of 4 and can be written as $\sqrt{4}$.

The table below contains the squares of the numbers from 1 to 15. These squares are called **perfect squares** since their square roots are whole numbers. Familiarity with these will help you to work problems in the next section.

Table of Perfect Squares		
$1^2 = 1$	$6^2 = 36$	$11^2 = 121$
$2^2 = 4$	$7^2 = 49$	$12^2 = 144$
$3^2 = 9$	$8^2 = 64$	$13^2 = 169$
$4^2 = 16$	$9^2 = 81$	$14^2 = 196$
$5^2 = 25$	$10^2 = 100$	$15^2 = 225$

The values on this table can be used to find either squares or square roots. For instance, $5^2 = 25$ or $\sqrt{25} = 5$.

EXAMPLE If $b^2 = 144$, what is the value of b?

Here the letter b stands for the square root of 144. In symbols, $b = \sqrt{144}$. To solve for b, find 144 in the right column above and identify its square root.

ANSWER: $b = 12$

 If you use a calculator, \boxed{C} $\boxed{1}$ $\boxed{4}$ $\boxed{4}$ $\boxed{\sqrt{}}$ $\boxed{=}$ $\boxed{\qquad 12.}$

Use the Table of Perfect Squares to find each value below.

1. $7^2 = $ _____

2. $10^2 = $ _____

3. $3^2 = $ _____

4. $15^2 = $ _____

5. $8^2 = $ _____

6. $4^2 = $ _____

7. $b^2 = 121$
 $b = $ _____

8. $c^2 = 25$
 $c = $ _____

9. $x^2 = 144$
 $x = $ _____

10. $a^2 = 1$
 $a = $ _____

11. $d = \sqrt{81}$
 $d = $ _____

12. $c = \sqrt{4}$
 $c = $ _____

13. $x = \sqrt{169}$
 $x = $ _____

14. $a = \sqrt{36}$
 $a = $ _____

15. $b = \sqrt{196}$
 $b = $ _____

Right Triangles and the Pythagorean Theorem

As you have seen, a right triangle has two sides that meet in a right angle. The side opposite the right angle is called the **hypotenuse** and is the longest side of a right triangle.

The Greek mathematician Pythagoras discovered an important relationship between the hypotenuse and the other two sides. He found that in a right triangle, the square of the hypotenuse is equal to the sum of the squares of the other two sides. We call this statement the **Pythagorean theorem.**

In symbols, using the labels on the triangle, you can write the Pythagorean theorem

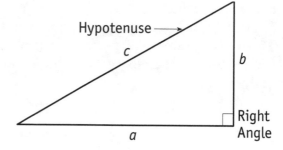

$$c^2 = a^2 + b^2$$

If you know the lengths of the other two sides, you can find the length of the hypotenuse by following these steps.

STEP 1 Find the square of the other two sides (often called legs) separately.

STEP 2 Add the squares of the sides found in Step 1.

STEP 3 Solve for the hypotenuse by finding the square root of the sum of the squares found in Step 2.

<u>EXAMPLE 1</u> What is the length of the hypotenuse of the triangle at the right?

STEP 1 Find the square of each side separately.

a. Substitute 6 for a and find a^2.

$$a^2 = 6^2 = 36$$

b. Substitute 8 for b and find b^2.

$$b^2 = 8^2 = 64$$

STEP 2 Add the squares found in Step 1.

$$a^2 + b^2 = 36 + 64 = 100$$

STEP 3 Since $c^2 = a^2 + b^2$, the sum of the squares found in Step 2 is equal to c^2. Then to find c, take the square root of 100.

$$c^2 = 100$$

$$c = \sqrt{100} = \mathbf{10}$$

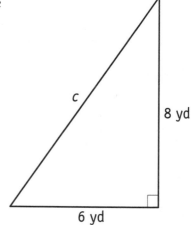

ANSWER: The length of the hypotenuse is 10 yards.

 Use the Pythagorean theorem to find the length of each hypotenuse in the triangles below. If you need to, refer to the Table of Perfect Squares on page 63, or use your calculator.

1.

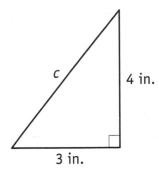

c 4 in.

3 in.

2.

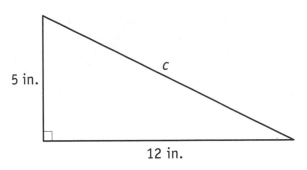

5 in. *c*

12 in.

3.

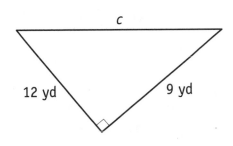

c

12 yd 9 yd

4.

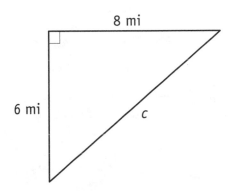

8 mi

6 mi *c*

5. What is the length of the hypotenuse of a right triangle whose legs are 5 feet and 12 feet long?

6. Right triangle CDE has sides measuring 4 meters and 3 meters. The hypotenuse is longer than either of these two sides. What is the length of the hypotenuse?

The Pythagorean theorem can also be used to find the length of a side if the lengths of the hypotenuse and one side are known.

EXAMPLE 2 What is the length of the unmeasured side of the triangle at the right?

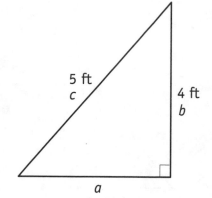

STEP 1 Find the square of the hypotenuse. Next, find the square of the given side.

a. Substitute 5 for c, and find c^2.

$$c^2 = 5^2 = 25$$

b. Substitute 4 for b, and find b^2.

$$b^2 = 4^2 = 16$$

Note: You could substitute 4 for a instead of b. A measured side can be called either a or b.

STEP 2 Write the Pythagorean theorem.

$$c^2 = a^2 + b^2$$

From Step 1, substitute 25 for c^2 and 16 for b^2.

$$25 = a^2 + 16$$

Notice that 25 is equal to a^2 plus 16.
To find a^2, subtract 16 from 25.

$$a^2 = 25 - 16$$
$$a^2 = 9$$

Note: To check that you have solved for a^2 correctly, notice that $9 + 16 = 25$.

STEP 3 Solve for a by taking the square root of 9.

$$a^2 = 9$$
$$\text{thus, } a = \sqrt{9} = 3$$

ANSWER: The length of unmeasured side a is 3 feet.

 Use the Pythagorean theorem to find the length of each unmeasured side below. Refer to the Table of Perfect Squares on page 63, or use your calculator.

1.

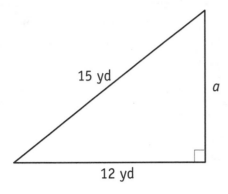

15 yd

12 yd

a

2.

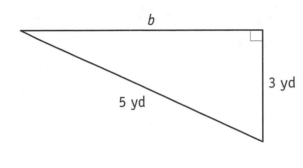

b

5 yd

3 yd

3.

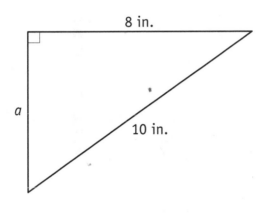

8 in.

a

10 in.

4.

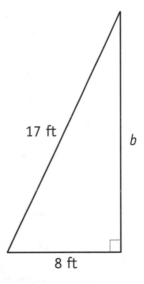

17 ft

8 ft

b

5. In right triangle ABC, the hypotenuse—side BC—is 13 inches long. Side AB is 5 inches long. What is the length of side AC?

6. Triangle QRS is a right triangle. Side QR, the hypotenuse, has a length of 17 centimeters. Side RS has a length of 15 centimeters. What is the length of side QS?

Note: Often, the sides of a right triangle have lengths of 3, 4, and 5 or a multiple of these, such as 6, 8, and 10. In these cases, the longest side, the hypotenuse, will be 5 or a multiple of 5. Problems 1, 2, and 3 illustrate this relationship.

Right Triangles and the Pythagorean Theorem: Applying Your Skills

The Pythagorean theorem is used to find the third side of a right triangle when the other two sides are known. In solving practical problems, be sure to correctly identify the hypotenuse as the side opposite the right angle.

Remember the Pythagorean theorem: $c^2 = a^2 + b^2$, where c is the hypotenuse.

EXAMPLE The support for a display sign attaches to the back of the sign 4 feet above the floor. The support rests on the floor 3 feet away. What is the length of the support? (The support is the hypotenuse of a right triangle formed by the sign and the floor.)

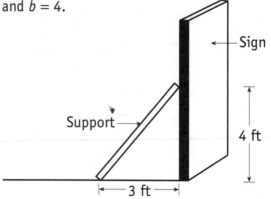

STEP 1 Find the square of each side. Let $a = 3$ and $b = 4$.

$$a^2 = 3^2 = 9 \text{ and } b^2 = 4^2 = 16$$

STEP 2 To find c^2, add a^2 and b^2.

$$c^2 = a^2 + b^2$$
$$c^2 = 9 + 16$$
$$\text{thus, } c^2 = 25$$

STEP 3 Take the square root.

$$c = \sqrt{25} = 5$$

ANSWER: The length of the support is 5 feet.

 Use the Pythagorean theorem to solve each problem.

1. A mover's ramp is placed on top of a 5-foot-high staircase. If the end of the ramp is on the floor 12 feet away, how long is the ramp?

2. A telephone pole support cable attaches to the pole 20 feet high. If the cable is 25 feet long, how far from the bottom of the pole does the cable attach to the ground?

3. A rectangle is a four-sided figure with four right angles. A diagonal divides a rectangle into two right triangles. What is the length of the diagonal in the rectangle below?

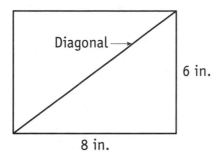

Diagonal →

6 in.

8 in.

4. George leaned a 17-foot ladder against the house. The bottom of the ladder is 8 feet from the house. How high is the top of the ladder?

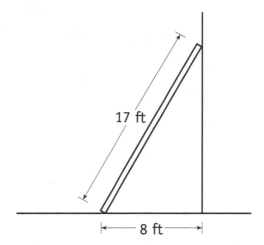

17 ft

8 ft

5. Joyce made a flower display in the front of her store. The display is in the shape of a right triangle. One side measures 5 feet. The second side measures 12 feet. How long is the side opposite the right angle?

12 ft

?

5 ft

6. Verdene is flying a model airplane at the end of the control line. The line is 17 yards long. The plane is above a point on the ground that is 15 yards away from Verdene. If the controls are held at a height of $1\frac{1}{2}$ yards, how high is the plane off the ground?
(**Hint:** height $= d + 1\frac{1}{2}$)

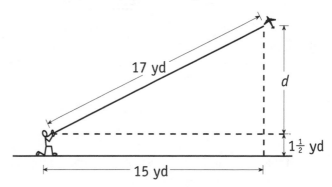

17 yd

d

15 yd

$1\frac{1}{2}$ yd

Triangles Review

This review covers the material you have just studied.

1. What is the measure of ∠R in the triangle below?

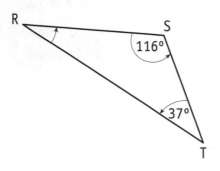

2. What is the missing angle Bill needs to find to cut his patio roof side piece?

3. The sum of two angles in a triangle is 138°. What is the value of the third angle?

In problems 4 and 5, circle the letter of the correct response.

4. The name given to triangle LMN is

 a. equilateral
 b. isosceles
 c. right
 d. scalene

5. The name given to triangle XYZ is

 a. equilateral
 b. isosceles
 c. right
 d. scalene

6. Solve for *d*.

 $$\frac{10}{17} = \frac{d}{51}$$

7. Triangles CDE and RST are similar triangles. What is the length of side CD in triangle CDE?

8. In triangle ABC, ∠A = 122° and ∠B = 30°.
 In triangle DEF, ∠D = 50° and ∠E = 122°.
 Is triangle ABC similar to triangle DEF?

9. To measure the distance across Blue River Canyon, Lucy measured
 the distances as shown on the drawing below. Use her drawing to
 determine the distance (d) across the canyon.

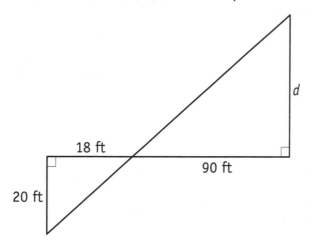

10. George is 6 feet tall and his
 shadow measured 8 feet long at
 noon. At the same time, a nearby
 flagpole cast a 40-foot shadow.
 What is the height of the
 flagpole?

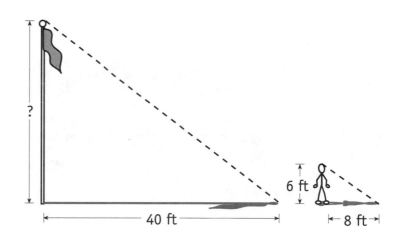

11. Find the square of each number.

 a. $2^2 =$ _____ c. $6^2 =$ _____

 b. $4^2 =$ _____ d. $8^2 =$ _____

12. Find the square root of each number.

 a. $\sqrt{9} =$ _____ c. $\sqrt{49} =$ _____

 b. $\sqrt{36} =$ _____ d. $\sqrt{81} =$ _____

13. What is the length of the hypotenuse of the triangle at the right?

14. A 13-foot ramp is attached to a 5-foot-high loading dock. How far is the bottom of the ramp from the dock?

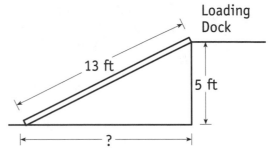

15. One side of a right triangle measures 8 feet. The other side measures 6 feet. What is the length of the hypotenuse?

Triangles Review Chart

Circle the number of any problem that you missed and review the appropriate pages. A passing score is 13 correct answers. If you miss more than three questions, you should review this chapter.

PROBLEM NUMBERS	SKILL AREA	PRACTICE PAGES
1, 3	sum of angles	44–45
2	sum of angles: applying skills	46–47
4, 5	definitions of triangles	48–50
6	similar triangles: proportion	54–57
7, 8	similar triangles	52–59
9, 10	similar triangles: applying skills	60–61
11, 12	squares and square roots	62–63
13, 15	Pythagorean theorem	64–67
14	Pythagorean theorem: applying skills	68–69

PLANE FIGURES

Plane Figures Skills Inventory

This inventory will let you know what you need to work on in the Plane Figures section of *Geometry*. Do all of the following problems that you can. On this inventory, you can use $\frac{22}{7}$ for π.

Add, subtract, multiply, or divide.

1. $\begin{array}{r} 5 \text{ ft } 9 \text{ in.} \\ 7 \text{ ft } 8 \text{ in.} \\ + 3 \text{ ft } 6 \text{ in.} \\ \hline \end{array}$

2. $\begin{array}{r} 8 \text{ m } 95 \text{ cm} \\ 2 \text{ m } 45 \text{ cm} \\ + 6 \text{ m } 75 \text{ cm} \\ \hline \end{array}$

3. $\begin{array}{r} 9 \text{ yd } 1 \text{ ft} \\ - 4 \text{ yd } 2 \text{ ft} \\ \hline \end{array}$

4. $\begin{array}{r} 3 \text{ km } 850 \text{ m} \\ - 1 \text{ km } 975 \text{ m} \\ \hline \end{array}$

5. $\begin{array}{r} 7 \text{ yd } 2 \text{ ft} \\ \times \quad\quad 9 \\ \hline \end{array}$

6. $\begin{array}{r} 5 \text{ cm } 4 \text{ mm} \\ \times \quad\quad 6 \\ \hline \end{array}$

7. $5\overline{)8 \text{ yd } 1 \text{ ft}}$

8. A blueprint calls for 175 feet of electrical wire. Is a roll containing 56 yards of wire long enough?

9. Harlan is installing a rain gutter along a 54-foot section of roof. He decides to do the job by using five sections of gutter placed end-to-end. If each section measures 12 feet 8 inches long, how much gutter will he have left over?

In problems 10 through 15, name each of the figures shown.

10. _____

11. _____

12. _____

13. _____

14. _____

15. _____

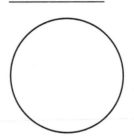

16. Match each formula on the left with the area it represents.

_____ 1. $\frac{1}{2}bh$ **a.** area of a trapezoid

_____ 2. πr^2 **b.** area of a rectangle

_____ 3. bh **c.** area of a square

_____ 4. lw **d.** area of a triangle

_____ 5. $\frac{1}{2}(b_1 + b_2)h$ **e.** area of a parallelogram

_____ 6. s^2 **f.** area of a circle

 In problems 17 through 22, find the perimeter and area of each figure. You may use your calculator.

17. P = _____

A = _____

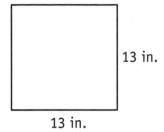

13 in.

13 in.

18. P = _____

A = _____

5 ft

9 ft

19. P = _____

A = _____

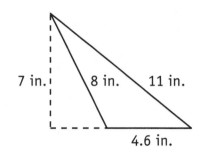

7 in. 8 in. 11 in.

4.6 in.

20. P = _____

A = _____

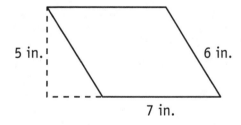

5 in. 6 in.

7 in.

21. P = _____

A = _____

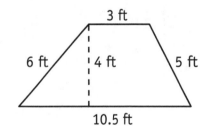

3 ft

6 ft 4 ft 5 ft

10.5 ft

22. C ≈ _____

A ≈ _____

14 in.

23. Edna's garden is in the shape of a rectangle having sides 35 meters and 21 meters. How much fencing material will it take to enclose the garden?

21 m

35 m

24. Bill has a circular swimming pool that has a radius of 7 feet. What is the approximate surface area of the pool?

7 ft

25. What is the floor area of the L-shaped room pictured below?

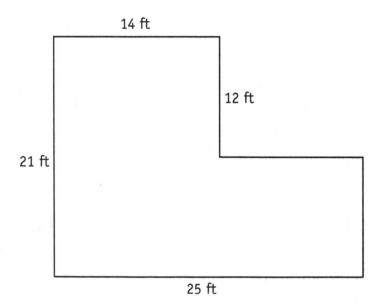

14 ft

12 ft

21 ft

25 ft

Plane Figures Inventory Chart

Use this inventory to see what you already know about plane figures and what you need to work on. A passing score is 21 correct answers. Even if you have a passing score, circle the number of any problem that you miss and turn to the practice pages indicated.

PROBLEM NUMBERS	SKILL AREA	PRACTICE PAGES
1, 2	addition: U.S. and metric units	82–83
3, 4	subtraction: U.S. and metric units	82–83
5, 6	multiplication: U.S. and metric units	84–85
7	division: U.S. units	84–85
8, 9	length units: applying skills	86–87
10, 11, 12, 13, 14, 15	recognizing shapes	90–91, 110
16	formulas	114
17	square: perimeter + area	94–97
18	rectangle: perimeter + area	98–101
19	triangle: perimeter + area	102–105
20	parallelogram: perimeter + area	106–109
21	trapezoid: perimeter + area	106–109
22	circle: circumference + area	110–113
23, 24	perimeter + area: applying skills	114–117
25	two-step area problem	118–121

Working with Customary Length Units

In the United States, we generally use the U.S. customary system of measurement. The familiar units of length are summarized below.

Unit	Abbreviation	Examples	Relation to Other Units
inch	in. or ″	5 in. or 5 ″	1 inch = $\frac{1}{12}$ foot
foot	ft or ′	7 ft or 7′	1 foot = 12 inches
			1 foot = $\frac{1}{3}$ yard
yard	yd	9 yd	1 yard = 36 inches
			1 yard = 3 feet
mile	mi	12 mi	1 mile = 5,280 feet
			1 mile = 1,760 yards

A length can be written in more than one way. Examples 1 and 2 show the use of multiplication to change a larger unit to a smaller unit.

EXAMPLE 1 Express 6 yards in feet.

STEP 1 Write the problem.
6 yards = _____ feet

STEP 2 You can substitute 3 feet for the word *yards* since 1 yard = 3 feet.
Multiply.
6(3 feet) = 18 feet

ANSWER: 18 feet

EXAMPLE 2 Express 7 feet in inches.

STEP 1 Write the problem.
7 feet = _____ inches

STEP 2 You can substitute 12 inches for the word *feet* since 1 foot = 12 inches.
Multiply.
7(12 in.) = 84 in.

ANSWER: 84 inches

Note: In geometry and algebra, multiplication is often indicated by parentheses (without the multiplication sign ×). When you see 6(3 feet), it means 6 × 3 feet.

Change each larger length unit to the smaller unit indicated. Each first problem is done as an example.

1. 8 ft = 8(12 in.) = **96 in.** 13 ft = _____ in. 4 ft = _____ in.

2. 7 yd = 7(3 ft) = **21 ft** 9 yd = _____ ft 16 yd = _____ ft

3. 3 mi = 3(5,280 ft) = **15,840 ft** 4 mi = _____ ft 2 mi = _____ ft

4. 2 mi = 2(1,760 yd) = **3,520 yd** 3 mi = _____ yd 5 mi = _____ yd

As shown in Examples 3 and 4, a length is often simplified by expressing it in the largest possible whole number length units.

EXAMPLE 3 Write 17 inches as feet and inches.

STEP 1 Write the problem.
17 in. = _____ ft _____ in.

STEP 2 Multiply to change inches to feet.
Since 1 inch = $\frac{1}{12}$ foot,
multiply 17 by $(\frac{1}{12})$.

$$17(\tfrac{1}{12}) = \tfrac{17}{12} = 1\tfrac{5}{12} \text{ ft}$$

Note: $\frac{5}{12}$ ft = 5 in.
Thus, $1\frac{5}{12}$ ft = **1 ft 5 in.**

ANSWER: 17 in. = 1 ft 5 in.

EXAMPLE 4 Simplify 4 feet 27 inches.

STEP 1 Change 27 inches to feet and inches by multiplying by $\frac{1}{12}$.

$$27 \text{ in.} = 27(\tfrac{1}{12}) = \tfrac{27}{12} \text{ ft}$$
$$= 2\tfrac{3}{12} \text{ ft}$$
$$= 2 \text{ ft } 3 \text{ in.}$$

STEP 2 Add 2 ft 3 in. to 4 ft.

$$\begin{array}{r} 4 \text{ ft} \\ + \ 2 \text{ ft } 3 \text{ in.} \\ \hline \mathbf{6 \text{ ft } 3 \text{ in.}} \end{array}$$

ANSWER: 4 ft 27 in. = 6 ft 3 in.

..

Simplify each length below. Each first problem is done as an example.

5. 21 in. = $21(\frac{1}{12}$ ft) =
$1\frac{9}{12}$ ft = **1 ft 9 in.**

19 in. = _____

31 in. = _____

6. 7 ft = $7(\frac{1}{3}$ yd) =
$2\frac{1}{3}$ yd = **2 yd 1 ft**

11 ft = _____

8 ft = _____

7. 1,900 yd = $1,900(\frac{1}{1,760}$ mi) =
$\frac{1,900}{1,760}$ mi = **1 mi 140 yd**

2,100 yd = _____

3,100 yd = _____

8. 5 ft 15 in. = 5 ft
 $\underline{+ \ 1 \text{ ft } 3 \text{ in.}}$
 6 ft 3 in.

6 ft 17 in. = _____

7 ft 23 in. = _____

9. 3 yd 7 ft = 3 yd
 $\underline{+ \ 2 \text{ yd } 1 \text{ ft}}$
 5 yd 1 ft

5 yd 8 ft = _____

11 yd 13 ft = _____

10. 1 mi 6,000 ft = 1 mi
 $\underline{+ \ 1 \text{ mi } 720 \text{ ft}}$
 2 mi 720 ft

1 mi 8,000 ft =

1 mi 10,280 ft =

Working with Metric Length Units

Most of the countries of the world use the **metric system** of measurement. This system is now being taught in many U.S. schools along with the familiar U.S. customary system.

The length units of the metric system are summarized below.

Unit	Abbreviation	Example	Relation to Other Units
millimeter	mm	8 mm	1 millimeter = 0.1 centimeter
centimeter	cm	11 cm	1 centimeter = 10 millimeters
meter	m	63 m	1 meter = 100 centimeters
kilometer	km	5 km	1 kilometer = 1,000 meters

To get a feeling for the size of metric units, compare each unit with a familiar U.S. unit.

The decimal (10, 100, 1,000) relation between units in the metric system makes it easier to work with metric units than with U.S. units.

For example, to change a length from a larger unit to a smaller unit, replace the larger unit with the equivalent number of smaller units and multiply.

EXAMPLES 2 cm = 2(10 mm) = 20 mm—since there are 10 mm in a centimeter.
5 m = 5(100 cm) = 500 cm—since there are 100 cm in a meter.
3 km = 3(1,000 m) = 3,000 m—since there are 1,000 m in a kilometer.

Change each larger length unit to the smaller unit indicated. Each first problem is done as an example.

1. 3 cm = 3(10 mm) 9 cm = _____ mm 4 cm = _____ mm 6 cm = _____ mm
 = **30 mm**

2. 8 m = 8(100 cm) 7 m = _____ cm 9 m = _____ cm 5 m = _____ cm
 = **800 cm**

3. 2 km = 2(1,000 m) 7 km = _____ m 10 km = _____ m 4 km = _____ m
 = **2,000 m**

As the following three rules show, simplifying lengths in the metric system is easier than in the U.S. customary system.

Rule 1 To simplify 10 or more millimeters, write the number in the 10's place as the number of centimeters.

EXAMPLE 14 millimeters = 1 centimeter 4 millimeters
↑——— number in 10's place

Rule 2 To simplify 100 or more centimeters, write the number in the 100's place as the number of meters.

EXAMPLE 346 centimeters = 3 meters 46 centimeters
↑——— number in 100's place

Rule 3 To simplify 1,000 or more meters, write the number to the left of the 100's place as the number of kilometers.

EXAMPLE 12,450 meters = 12 kilometers 450 meters
↑——— number to the left of 100's place

Use these rules to simplify more complicated lengths.

EXAMPLE 1 Simplify 7 cm 29 mm.

 STEP 1 Write 29 mm as cm and mm.
 Use Rule 1.

 29 mm = 2 cm 9 mm

 STEP 2 Add 2 cm 9 mm to 7 cm.
 7 cm
 + 2 cm 9 mm
 9 cm 9 mm

ANSWER: 7 cm 29 mm = 9 cm 9 mm

EXAMPLE 2 Simplify 9 m 368 cm.

 STEP 1 Write 368 cm as m and cm.
 Use Rule 2.

 368 cm = 3 m 68 cm

 STEP 2 Add 3 m 68 cm to 9 m.
 9 m
 + 3 m 68 cm
 12 m 68 cm

ANSWER: 9 m 368 cm = 12 m 68 cm

Simplify each length below. Each first problem is done as an example.

4. 28 mm = **2 cm 8 mm** 79 mm = _____ 83 mm = _____

5. 426 cm = **4 m 26 cm** 748 cm = _____ 104 cm = _____

6. 5 cm 19 mm = 5 cm 7 cm 57 mm = _____ 9 cm 39 mm = _____
 + 1 cm 9 mm
 6 cm 9 mm

7. 8 m 596 cm = 8 m 5 m 856 cm = _____ 12 m 800 cm = _____
 + 5 m 96 cm
 13 m 96 cm

Addition and Subtraction of Length Units

To add two or more lengths, add the number of each length unit separately and then simplify the answer. An example is shown below both for lengths in the U.S. customary system and lengths in the metric system.

CUSTOMARY SYSTEM

$$
\begin{array}{r}
6 \text{ ft } 8 \text{ in.} \\
3 \text{ ft } 6 \text{ in.} \\
+ \ 4 \text{ ft } 3 \text{ in.} \\
\hline
13 \text{ ft } 17 \text{ in.}
\end{array}
$$

ANSWER: 14 ft 5 in.
(Since 17 in. = 1 ft 5 in.)

Addition
STEP 1 Add each column separately.
STEP 2 Simplify the sum.

METRIC SYSTEM

$$
\begin{array}{r}
7 \text{ cm } \ 9 \text{ mm} \\
4 \text{ cm } \ 4 \text{ mm} \\
+ \ 5 \text{ cm } \ 8 \text{ mm} \\
\hline
16 \text{ cm } 21 \text{ mm}
\end{array}
$$

ANSWER: 18 cm 1 mm
(Since 21 mm = 2 cm 1 mm)

To subtract one length from another, subtract each length unit separately, starting with the unit in the right-hand column. Borrowing is often necessary in subtraction and is illustrated in both the U.S. and the metric examples below.

Subtract:
$$
\begin{array}{r}
5 \text{ ft } 2 \text{ in.} \\
- 3 \text{ ft } 9 \text{ in.}
\end{array}
$$

Borrow: 1 ft = 12 in.
(12 in. + 2 in. = 14 in.)

Write:
$$
\begin{array}{r}
\overset{4}{\cancel{5}} \text{ ft } \overset{14}{\cancel{2}} \text{ in.} \\
- 3 \text{ ft } 9 \text{ in.}
\end{array}
$$
ANSWER: 1 ft 5 in.

Subtraction
STEP 1 If necessary, borrow 1 whole unit from the left column. Add this borrowed unit to the right column.
STEP 2 Subtract each column.

Subtract:
$$
\begin{array}{r}
8 \text{ cm } 5 \text{ mm} \\
- 4 \text{ cm } 8 \text{ mm}
\end{array}
$$

Borrow: 1 cm = 10 mm
(10 mm + 5 mm = 15 mm)

Write:
$$
\begin{array}{r}
\overset{7}{\cancel{8}} \text{ cm } \overset{15}{\cancel{5}} \text{ mm} \\
- 4 \text{ cm } 8 \text{ mm}
\end{array}
$$
ANSWER: 3 cm 7 mm

Add the following lengths and simplify each answer. Each first problem is done as an example.

1.
 5 ft 7 in.
 + 3 ft 9 in.
 8 ft 16 in.
 = **9 ft 4 in.**

 7 ft 8 in.
 + 1 ft 5 in.

 4 yd 1 ft
 + 2 yd 2 ft

 8 mi 400 yd
 + 3 mi 1700 yd

2.
 4 yd 1 ft
 2 yd 2 ft
 + 1 yd 2 ft
 7 yd 5 ft
 = **8 yd 2 ft**

 5 yd 2 ft
 2 yd 1 ft
 + 4 yd 2 ft

 7 ft 8 in.
 3 ft 9 in.
 + 5 ft 3 in.

 2 yd 2 ft
 3 yd 1 ft
 + 6 yd

3.
 5 cm 9 mm
 + 1 cm 2 mm
 6 cm 11 mm
 = **7 cm 1 mm**

 6 cm 8 mm
 + 4 cm 6 mm

 9 m 58 cm
 + 1 m 90 cm

 2 km 460 m
 + 3 km 750 m

4.
 9 m 57 cm
 7 m 40 cm
 + 3 m 60 cm
 19 m 157 cm
 = **20 m 57 cm**

 12 m 75 cm
 9 m 80 cm
 + 6 m 35 cm

 25 cm 8 mm
 11 cm 4 mm
 + 8 cm 5 mm

 5 km 100 m
 3 km 500 m
 + 2 km 750 m

Subtract the following lengths. Each first problem is done as an example.

5.
 $^5\cancel{6}$ ft $\overset{17}{\cancel{7}}$ in.
 − 3 ft 9 in.
 2 ft 8 in.

 7 ft 3 in.
 − 4 ft 8 in.

 3 yd 1 ft
 − 1 yd 2 ft

 5 ft 7 in.
 − 2 ft 11 in.

6.
 $^3\cancel{4}$ yd $\overset{4}{\cancel{1}}$ ft
 − 2 yd 2 ft
 1 yd 2 ft

 6 yd 1 ft
 − 1 yd 2 ft

 3 ft 2 in.
 − 1 ft 9 in.

 7 yd 1 ft
 − 6 yd 2 ft

7.
 $^6\cancel{7}$ cm $\overset{18}{\cancel{8}}$ mm
 − 3 cm 9 mm
 3 cm 9 mm

 12 cm 7 mm
 − 4 cm 8 mm

 35 m 50 cm
 − 17 m 85 cm

 9 km 150 m
 − 3 km 500 m

Multiplication and Division of Length Units

To multiply a length by a number, multiply each length unit separately. Then, simplify the answer when possible. The examples below show how this is done.

CUSTOMARY SYSTEM

$$\begin{array}{r} 3 \text{ yd} \quad 2 \text{ ft} \\ \times \qquad 6 \\ \hline 18 \text{ yd} \; 12 \text{ ft} \end{array}$$

ANSWER: 22 yd
(Since 12 ft = 4 yd,
18 yd + 4 yd = **22 yd**.)

Multiplication
STEP 1 Multiply each column.
STEP 2 Simplify the product.

METRIC SYSTEM

$$\begin{array}{r} 8 \text{ m} \quad 70 \text{ cm} \\ \times \qquad 3 \\ \hline 24 \text{ m} \; 210 \text{ cm} \end{array}$$

ANSWER: 26m 10 cm
(Since 210 cm = 2 m 10 cm,
24 m + 2 m 10 cm = **26 m 10 cm**.)

To divide a length by a number, first divide the number into the largest length unit. Next, change any remainder into the next smallest unit in the problem. Add the remainder to the number of small units in the problem, and then divide this sum by the number. The examples below demonstrate this process.

Divide:

$$\begin{array}{r} 1 \text{ ft} \; 5 \text{ in.} \\ 4 \overline{) 5 \text{ ft} \; 8 \text{ in.}} \\ -4 \\ \hline \end{array}$$

Remainder: ⅟ ft 8 in.
Borrow and add: +12 in.
(1 ft = 12 in.) 20 in.
Divide 20 by 4: $\begin{array}{r} -20 \\ \hline 0 \end{array}$

ANSWER: 1 ft 5 in.

Division
STEP 1 Divide the first column.
STEP 2 Change the remainder of the first column to the units of the second column. Add this remainder to the second column.
STEP 3 Add the numbers in the second column. Divide this sum.

Divide:

$$\begin{array}{r} 2 \text{ m} \; 64 \text{ cm} \\ 3 \overline{) 7 \text{ m} \; 92 \text{ cm}} \\ -6 \\ \hline \end{array}$$

Remainder: ⅟ m 92 cm
Borrow and add: +100 cm
(1 m = 100 cm) 192 cm
Divide 192 by 3: $\begin{array}{r} -192 \\ \hline 0 \end{array}$

ANSWER: 2 m 64 cm

Multiply the following lengths and simplify each answer. Each first problem is done as an example.

1.
$$\begin{array}{r} 6\ \text{ft}\quad 9\ \text{in.} \\ \times\qquad 3 \\ \hline 18\ \text{ft}\ 27\ \text{in.} \\ = \textbf{20 ft 3 in.} \end{array}$$

$$\begin{array}{r} 7\ \text{ft}\ 6\ \text{in.} \\ \times\qquad 8 \\ \hline \end{array}$$

$$\begin{array}{r} 9\ \text{yd}\ 2\ \text{ft} \\ \times\qquad 4 \\ \hline \end{array}$$

$$\begin{array}{r} 2\ \text{mi}\ 400\ \text{yd} \\ \times\qquad 7 \\ \hline \end{array}$$

2.
$$\begin{array}{r} 5\ \text{yd}\quad 2\ \text{ft} \\ \times\qquad 7 \\ \hline 35\ \text{yd}\ 14\ \text{ft} \\ = \textbf{39 yd 2 ft} \end{array}$$

$$\begin{array}{r} 9\ \text{yd}\ 1\ \text{ft} \\ \times\qquad 8 \\ \hline \end{array}$$

$$\begin{array}{r} 4\ \text{ft}\ 11\ \text{in.} \\ \times\qquad 6 \\ \hline \end{array}$$

$$\begin{array}{r} 2\ \text{yd}\ 2\ \text{ft} \\ \times\qquad 9 \\ \hline \end{array}$$

3.
$$\begin{array}{r} 8\ \text{cm}\quad 9\ \text{mm} \\ \times\qquad 6 \\ \hline 48\ \text{cm}\ 54\ \text{mm} \\ = \textbf{53 cm 4 mm} \end{array}$$

$$\begin{array}{r} 9\ \text{cm}\ 4\ \text{mm} \\ \times\qquad 7 \\ \hline \end{array}$$

$$\begin{array}{r} 3\ \text{cm}\ 8\ \text{mm} \\ \times\qquad 5 \\ \hline \end{array}$$

$$\begin{array}{r} 7\ \text{m}\ 45\ \text{cm} \\ \times\qquad 4 \\ \hline \end{array}$$

Divide each length. Each first problem is done as an example.

4.
$$\begin{array}{r} \textbf{1 ft}\quad \textbf{11 in.} \\ 5\overline{)9\ \text{ft}\quad 7\ \text{in.}} \\ \underline{5\qquad\qquad} \\ \cancel{4}\ \text{ft} = 48\ \text{in.} \\ 55\ \text{in.} \\ \underline{55\qquad} \end{array}$$

$$4\overline{)9\ \text{ft}\ 8\ \text{in.}}$$

$$9\overline{)21\ \text{ft}\ 9\ \text{in.}}$$

$$6\overline{)13\ \text{ft}\ 6\ \text{in.}}$$

5.
$$\begin{array}{r} \textbf{1 yd 1 ft} \\ 7\overline{)9\ \text{yd}\ 1\ \text{ft}} \\ \underline{7\qquad\quad} \\ \cancel{2}\ \text{yd} = 6\ \text{ft} \\ 7\ \text{ft} \\ \underline{7\quad} \end{array}$$

$$8\overline{)26\ \text{yd}\ 2\ \text{ft}}$$

$$2\overline{)9\ \text{yd}\ 1\ \text{ft}}$$

$$5\overline{)48\ \text{yd}\ 1\ \text{ft}}$$

6.
$$\begin{array}{r} \textbf{2 cm 3 mm} \\ 6\overline{)13\ \text{cm}\ 8\ \text{mm}} \\ \underline{12\qquad\qquad} \\ \cancel{1}\ \text{cm} = 10\ \text{mm} \\ 18\ \text{mm} \\ \underline{18\qquad} \end{array}$$

$$9\overline{)24\ \text{cm}\ 3\ \text{mm}}$$

$$4\overline{)7\ \text{m}\ 20\ \text{cm}}$$

$$7\overline{)8\ \text{km}\ 820\ \text{m}}$$

Working with Length Units: Applying Your Skills

Solve each problem.

1. Bill needs 109 inches of baseboard to place along the family room floor. Will a board that is 8 feet long be enough?

2. To fence her backyard, Heather estimated it would take 32 yards of fencing material. Will two rolls of fencing be enough if each roll contains a 50-foot length of material?

3. To make a long garden hose, Leona connected three shorter hoses together. If each of the shorter hoses measures 24 feet 9 inches, what is the length of the combined hose?

4. Daniel uses gold wire in his jewelry-making business. Of his supply, Daniel has used all but four separate lengths of the precious wire. These lengths measure 29 cm 9 mm, 11 cm 4 mm, 14 cm 7 mm, and 16 cm 9 mm. If the wire costs $20 per centimeter (or $2 per millimeter), what is the value of the wire that Daniel has left?

5. On Mona's thirteenth birthday, she measured 4 feet 9 inches tall. How much has she grown since her sixth birthday when her height was 3 feet 11 inches?

6. When Billy was in the 7th grade, he could throw a football 34 yards 2 feet. Now, in the 12th grade, he can throw it 47 yards 1 foot. How much farther can he throw it now than when he was in the 7th grade?

7. Terry's height is 176 centimeters. His friend, Fred, stands 2 meters 3 centimeters tall. Which boy is taller?

8. To frame a front door, Susy needs three pieces of molding: two pieces measuring 6′ 8″ each, and one piece measuring 3′ 2″. If she cuts these pieces from a board that measures 17′ 2″, how much molding will she have left?

9. Chris entered a 100-kilometer bicycle race. During the first three hours, she rode the following distances: 1st hour = 23 km 800 m, 2nd hour = 25 km 350 m, 3rd hour = 26 km 575 m. After the third hour, how much farther did she still have to ride?

10. Jamie designed a picnic table that requires six 2″ by 6″ boards, each board measuring 6′ 9″ long. What is the total length of 2″ by 6″ boards needed?

11. A quilt pattern calls for four pieces of fabric, each measuring 7 inches long, and four pieces of fabric, each measuring 8 inches long. How much fabric will be left over if all these pieces are cut from a piece that is 6 feet 5 inches long?

12. Barbara has 25 decorative bricks. Each brick is 14 inches long. If the bricks are placed end-to-end in a straight line, what distance will they cover? Give your answer to the nearest foot.

13. If Vince wants to cut an 8 foot 9 inch pipe into five equal pieces, how long should he cut each piece?

14. Shauna is going to divide a 10 meter long ribbon into four equal lengths. To the nearest centimeter, how long will each smaller ribbon be?

Perimeter

The distance around a flat (plane) object is known as its **perimeter.**
For example, the distance around a field or a lake is called its perimeter.
The symbol for perimeter is *P.*

Perimeter is measured in length units. To find the perimeter of a
many-sided figure, add the lengths of the sides.

EXAMPLE 1 What is the perimeter of the figure at the right?

To find the perimeter, add the lengths of the five sides.

$$
\begin{array}{r}
6 \text{ ft} \\
3 \text{ ft} \\
7 \text{ ft} \\
2 \text{ ft} \\
+ \, 4 \text{ ft} \\
\hline
P = \mathbf{22 \text{ ft}}
\end{array}
$$

ANSWER: The perimeter is 22 ft or 7 yd 1 ft

EXAMPLE 2 Find the perimeter of the field at the right.

STEP 1 Add the lengths of the four sides.

$$
\begin{array}{r}
18 \text{ yd } 2 \text{ ft} \\
21 \text{ yd } 2 \text{ ft} \\
19 \text{ yd } 1 \text{ ft} \\
+ \, 20 \text{ yd } 2 \text{ ft} \\
\hline
78 \text{ yd } 7 \text{ ft}
\end{array}
$$

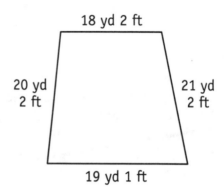

STEP 2 Simplify the answer.
a. Change 7 ft to 2 yd 1 ft.
b. Add 2 yd 1 ft to 78 yd.

$$
\begin{array}{r}
78 \text{ yd} \\
+ \quad 2 \text{ yd } 1 \text{ ft} \\
\hline
\mathbf{80 \text{ yd } 1 \text{ ft}}
\end{array}
$$

ANSWER: The perimeter is 80 yd 1 ft.

Find the perimeter of each figure.

1. *P* = _____

9 in.
14 in.
7 in.
10 in.

2. *P* = _____

11 yd
5 yd
8 yd

3. $P =$ _____

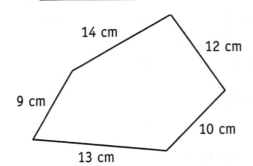

14 cm

12 cm

9 cm

10 cm

13 cm

4. $P =$ _____

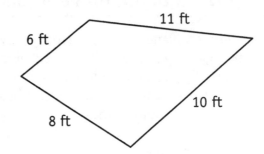

11 ft

6 ft

8 ft

10 ft

Find the perimeter of each figure below and simplify your answer.

5. $P =$ _____

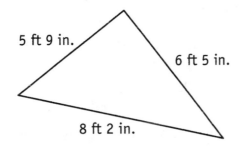

5 ft 9 in.

6 ft 5 in.

8 ft 2 in.

6. $P =$ _____

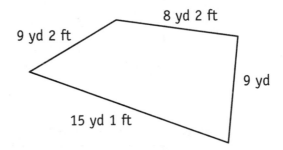

8 yd 2 ft

9 yd 2 ft

9 yd

15 yd 1 ft

7. $P =$ _____

2 m

1 m 7 cm

1 m 90 cm

3 m 60 cm

2 m 20 cm

8. $P =$ _____

4 mi 2,000 ft

2 mi

4 mi 2,700 ft

5 mi 1,500 ft

Recognizing Common Polygons

A **polygon** is a plane (flat) figure formed by three or more lines. A **regular polygon** is a polygon that has equal sides and equal angles. Several types of polygons are given special names.

Below are listed the names and descriptions of five polygons with which you should become familiar.

Name	Example	Description
Triangle	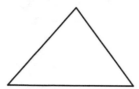	A **triangle** is a polygon formed by three lines.
Square	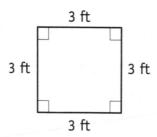	A **square** is a regular polygon with four equal sides, two pairs of parallel sides, and four right angles.
Rectangle	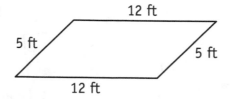	A **rectangle** is a four-sided polygon with two pairs of parallel sides and four right angles. The opposite sides of a rectangle have equal lengths.
Parallelogram		A **parallelogram** is a four-sided polygon with two pairs of parallel sides. The opposite sides are equal, and the opposite angles are equal. If all four sides of a parallelogram are equal, the parallelogram is called a **rhombus.**
Trapezoid		A **trapezoid** is a four-sided polygon with one pair of parallel sides called **bases.** All four sides of a trapezoid can have different lengths.

Label each figure below with its polygon name: triangle, square, rectangle, parallelogram, or trapezoid.

1. _____

2. _____

3. _____

SPEED ZONE AHEAD

4. _____

5. _____

6. _____

7. _____

DINO's
Restaurant
& Lounge

Sign

8. _____

Map of Johnson's Property

9. _____

Chimney

10. _____

Wedge

11. _____

Contemporary's
NUMBER POWER
Geometry
ROBERT MITCHELL & DONALD PRICKEL

4

12. _____

Toy Block

Area

Area is a measure of surface. A larger surface has a larger area. For example, the total floor area of a house is larger than the area of any one of its rooms. The symbol for area is *A*.

To measure area, you can use an **area unit** in the shape of a square.

For example, you can use this square [] as your area unit.

The surface area of a figure can be measured by the number of square area units that fit inside the figure. One way to find the area is to divide a figure into square area units and then count the units that fit inside the figure.

EXAMPLE What is the area of the rectangle at the right?

STEP 1 Divide the rectangle into square area units.

STEP 2 To find the area of the rectangle, count the number of area units that fit inside the rectangle.

ANSWER: *A* = 6 area units

What is the area of each figure below? Each figure has been divided into square area units.

1. *A* = _____ area units

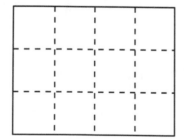

2. *A* = _____ area units

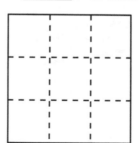

3. *A* = _____ area units

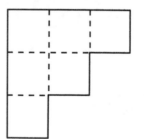

4. *A* = _____ area units

5. *A* = _____ area units

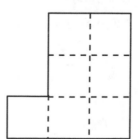

6. *A* = _____ area units

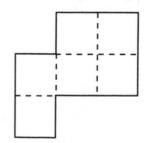

Not all figures can be simply divided into a whole number of square area units. Find the area of each figure below as accurately as possible. Some answers will include fractions.

7. $A =$ _____ area units

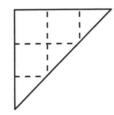

8. $A =$ _____ area units

9. $A =$ _____ area units

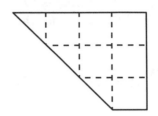

10. $A =$ _____ area units

11. $A =$ _____ area units

12. $A =$ _____ area units

For many curved figures, counting area units gives only an approximate area. See how accurate you can be in finding the area of the two curved surfaces below.

13. $A =$ _____ area units

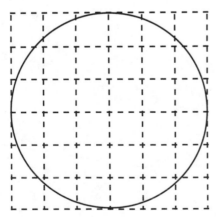

14. $A =$ _____ area units

Working with Squares

In the pages ahead, you'll learn to use **formulas** to find perimeters and areas of common geometrical figures. As you saw in the chapter on triangles, a formula is a rule that uses letters in place of numbers. To use a formula, you replace the letters with numbers and then do the arithmetic.

PERIMETER

The sides of a **square** are all the same length. Because of this, you can write a simple formula for finding the perimeter of a square.

$$P = 4s$$

This means that perimeter equals four times s where s is the length of each side.

This formula can be rearranged to give the length of each side when you know the perimeter.

$$s = \frac{P}{4}$$

This means that a side's length is found by dividing the perimeter by four.

EXAMPLE 1 What is the perimeter of a square with sides 3 feet long?
Substitute 3 for s in the formula $P = 4s$

$$P = 4 \times 3 = 12$$

ANSWER: 12 ft or 4 yd

EXAMPLE 2 What is the length of each side of a square that has a perimeter of 140 inches?

Substitute 140 for P in the formula $s = \frac{P}{4}$.

$$s = \frac{140}{4} = 35$$

ANSWER: 35 in. or 2 ft 11 in.

 Use the formula *P = 4s* to find the perimeter of each square below.

1. *P =* _____

7 in.

2. *P =* _____

9 yd

3. *P =* _____

12 ft 5 in.

4. *P =* _____

21 m

5. *P =* _____

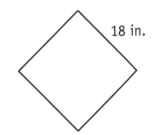

18 in.

6. *P =* _____

5 cm
7 mm

In problems 7 through 12, find the perimeter or length of one side as requested. Notice that words such as *enclose, surround, wrap,* and *around* indicate that you are to solve for the perimeter.

 7. Michelle drives her remote-control car around a square track. Each side of the track is 30 feet long. How many times does the car need to go around the track to go a total distance of 1,200 feet?

8. Jane wants to make a border for a tablecloth with lace material. How much material will she need if the tablecloth measures 4 feet 4 inches by 4 feet 4 inches?

 9. How much fence material should Bill buy to surround a garden space that is in the shape of a square 17 yards on a side?

10. Allowing 3 feet for the door, how much baseboard would John need to enclose a square room where each wall measures 11 feet?

 11. If the perimeter of a square field measures 440 yards, what is the length of each side?

12. The distance around a square room measures 101 feet 7 inches. Estimate the length of each side of the room.

AREA

The area of common geometrical figures is usually found by using an **area formula.** To use an area formula, substitute numbers for letters and then multiply or divide as indicated. Label your answer with appropriate area units.

Common area units are the following:

U.S. Units	Abbreviation
square inch	sq in.
square foot	sq ft
square yard	sq yd
square mile	sq mi

Metric Units	Abbreviation
square millimeter	mm²
square centimeter	cm²
square meter	m²
square kilometer	km²

AREA OF A SQUARE

The area of a **square** is given by the formula

$$A = s^2$$

where s stands for side and s^2 means $s \times s$.

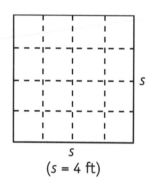
s
$(s = 4 \text{ ft})$

The drawing at the right illustrates the use of this formula to find the area of a square. The formula means that the area of the square is found by *squaring* a side—multiplying the value of a side times itself. For the square shown here, that means

$$A = 4 \times 4 = 16.$$

Count the number of smaller squares within the larger one. Does this also total 16?

If the area of a square is known, the length of the side can easily be found. The side is equal to the square root of the area:

$$s = \sqrt{A}$$

Try this formula on the square at the right. Notice that the area is 4 square inches. Do you see that the length of each side is 2 inches? ($\sqrt{4} = 2$)

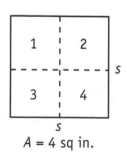
| 1 | 2 |
| 3 | 4 |
s
$A = 4 \text{ sq in.}$

<u>EXAMPLE 1</u> What is the area of a square that has a side length of 12 meters?

Replace s by 12 in the area formula.

$$A = s^2 = s \times s$$
$$= 12 \times 12 = \mathbf{144}$$

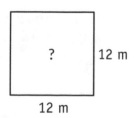
? 12 m
12 m

ANSWER: 144 m²

<u>EXAMPLE 2</u> What is the length of the side of a square that has an area
of 36 square feet?

To find the length of the side take the square root of 36.

$$s = \sqrt{36} = 6$$

ANSWER: 6 feet

36 sq ft ?

?

..

In problems 1 through 3, use the formula $A = s^2$ or $s = \sqrt{A}$ to find the
area or the length of the side of each square. Be sure that area answers
are given in square units.

1. $A =$ _____

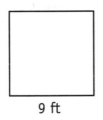

9 ft

2. $s =$ _____

$A = 64$ cm²

3. $A =$ _____

4.5 in.

In problems 4 through 8, find the area or the length of a side as
requested. Notice that words such as *covering, surface,* and *space* are
often used to indicate area.

4. Bill is covering a square table with glass. How much glass will be
needed if the table measures 30 centimeters on each side?

5. How much tile will be needed to cover a square utility room that is
8 feet on each side?

6. What is the length of each side of a square tablecloth that covers a
surface area of 9 square feet?

7. Find the area of a square garden that is 17 yards on a side.

8. A square table is to be covered with small square ceramic tiles. The
table has an area of 4 square feet. How many tiles are needed if each
tile measures 3 inches on one side?

Working with Rectangles

PERIMETER

The perimeter of a **rectangle** is found by adding the four sides

$$P = l + w + l + w$$

where *l* stands for length and *w* stands for width.

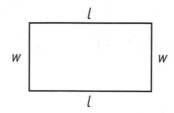

This formula is most often simplified by combining the *l*'s and combining the *w*'s.

$$P = 2l + 2w$$

This means that the perimeter is equal to twice the length plus twice the width.

> **Note:** When multiplication and addition occur in the same problem, do the multiplication first.

<u>EXAMPLE</u> What is the perimeter of a rectangle that has a length of 8 feet and a width of 5 feet?

Substitute 8 for *l* and 5 for *w* in the perimeter formula.

$$P = 2l + 2w$$
$$P = (2 \times 8) + (2 \times 5)$$
$$= 16 + 10$$
$$= \mathbf{26}$$

ANSWER: The perimeter is 26 ft or 8 yd 2 ft.

Use the formula $P = 2l + 2w$ to find the perimeter of each rectangle.

1. $P =$ _____

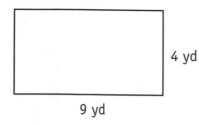

9 yd (bottom), 4 yd (right)

2. $P =$ _____

21 ft (right), 13 ft (bottom)

3. $P =$ _____

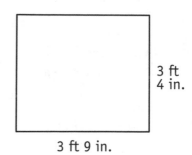

3 ft 4 in. (right), 3 ft 9 in. (bottom)

4. P = _____

12 m
85 cm

7 m 60 cm

5. P = _____

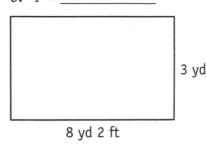

3 yd

8 yd 2 ft

6. P = _____

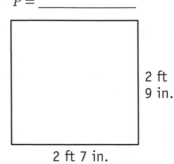

2 ft
9 in.

2 ft 7 in.

7. How many feet of fencing material will Louise need to surround her garden that measures 15 feet 8 inches on two sides and 19 feet 10 inches on the other two sides?

 8. After adding 4 extra inches for each corner, how much molding will Jim need to enclose a window that measures 36 inches by 40 inches?

36 in.

40 in.

9. Julie plans to add a lace edge to her bedspread which is in the shape of a rectangle. How much lace will she need if the bedspread measures 6 feet by 5 feet 8 inches?

 10. A rectangular field measures 100 yards by 76 yards. How many laps around the field does Len have to run in order to run one mile? (1 mile = 1,760 yards)

76 yd

100 yd

11. Determine the perimeter of a lawn which is in the shape of a rectangle and measures 35 meters 50 centimeters by 48 meters 25 centimeters?

12. Subtracting 2 feet 11 inches for door space, how much baseboard would be needed to go around a rectangular room that measures 11 feet 4 inches by 12 feet 7 inches?

AREA

The area of a **rectangle** is given by the formula

$$A = lw$$

where l stands for length and w stands for width.
(Two letters standing side-by-side indicate multiplication.)

The rectangle above is an illustration of the use of the area formula. The area is easily found by counting the area units (12) or by multiplying the length (4) by the width (3): $4 \times 3 = 12$.

The area formula can be rewritten so that a side can be found if the area is given. Because you multiply the sides to find the area, you can divide the area by one side to find the second side. The formula for finding the length or the width is either

$$5 \overline{)8} \text{ yd 1 ft}$$

These formulas show that one side is equal to the area divided by the second side.

EXAMPLE 1 What is the area of a rectangle that has a length of 5 feet and width of 4 feet?

To find the area, multiply the length (5) by the width (4).

$$A = 5 \times 4 = \mathbf{20}$$

ANSWER: 20 square feet

EXAMPLE 2 What is the width of a rectangle that has an area of 48 square yards and a length of 12 yards?

To find the width, divide the area (48) by the length (12).

$$\frac{1}{12}$$

ANSWER: 4 yards

..

: **Note:** Area is given in square units but a side is not.

..

 Use the formula $A = lw$ or $\frac{1}{3}$ to find the area, length, or width of each rectangle.

1. $A = $ _____

9 ft

13 ft

2. $A = $ _____

$3\frac{1}{2}$ km

2 km

3. $w = $ _____

$A = 136$ sq in.

17 in.

4. $A = $ _____

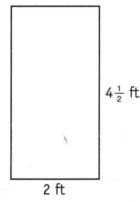

$4\frac{1}{2}$ ft

2 ft

5. $l = $ _____

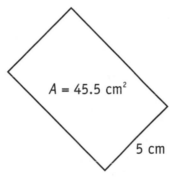

$A = 45.5$ cm²

5 cm

6. $A = $ _____

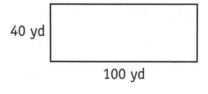

40 yd

100 yd

7. Brenda found that a bedroom floor measures 5 yards long and $4\frac{1}{2}$ yards wide. How many square yards of carpet would she need to cover this floor?

8. How many square feet of tile are needed to put a new tile surface on an outdoor patio that measures 32 feet by 20 feet?

9. The label on a gallon of interior wall paint reads: "Coverage: 400 square feet." If a wall is 8 feet high, what maximum length of wall could be painted by $1\frac{1}{2}$ gallons?

10. A standard piece of typing paper measures 11 inches long by $8\frac{1}{2}$ inches wide. Is the area of a piece of this paper larger than 100 square inches?

11. A window measures $2\frac{1}{4}$ feet high and 4 feet wide. How many square feet of glass will be needed to replace the broken window?

12. Estimate the width of a field that has an area of 6,000 square meters and length of 118.5 meters.

Working with Triangles

PERIMETER

The perimeter of a **triangle** is found by adding its three sides.

$$P = s_1 + s_2 + s_3$$

where s_1 stands for side 1,
s_2 stands for side 2, and
s_3 stands for side 3.

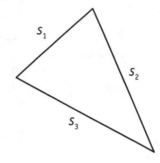

EXAMPLE What is the perimeter of a triangle that measures 5 feet by 3 feet by 7 feet?

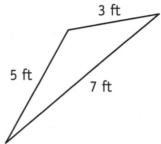

To find the perimeter, 5
add the three sides. 3
 + 7
 ‾‾‾‾
 15

ANSWER: The perimeter is 15 feet or 5 yards.

Find the perimeter of each triangle.

1. $P = $ _____

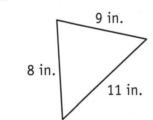

9 in.

8 in.

11 in.

2. $P = $ _____

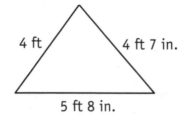

4 ft 4 ft 7 in.

5 ft 8 in.

3. $P = $ _____

11 cm 6 mm

9 cm 7 cm
7 mm 1 mm

4. $P = $ _____

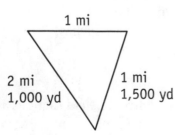

1 mi

2 mi 1 mi
1,000 yd 1,500 yd

5. $P = $ _____

14 m

12 m
65 cm

4 m
80 cm

6. $P = $ _____

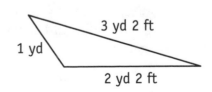

3 yd 2 ft

1 yd

2 yd 2 ft

38 yd

18 yd

27 yd

7. A garden in the shape of a triangle has sides with lengths 18 yards, 27 yards, and 38 yards. How much fencing material will it take to enclose it?

8. A roof gable is in the shape of a triangle. If each of two sides is 18 feet 6 inches and the third side is 38 feet 8 inches, what is the perimeter of the gable?

9. What is the distance around a triangular lot that measures 126 meters by 236 meters by 202 meters?

10. A triangular mirror has three equal sides. Each side measures 8 inches. Allowing three extra inches for each corner, how much molding will be needed to enclose it?

8 in. 8 in.

8 in.

11. A jogging trail has sides that measure 168 yards 2 feet, 175 yards, and 96 yards 1 foot. How many laps around the trail is equal to 1 mile (1,760 yards)?

12. A triangular piece of cut glass is to be enclosed in a solder strip. What length of solder is needed if the glass has sides measuring 4 cm 3 mm, 5 cm 6 mm, and 13 cm 3 mm?

13. Joe's Deli has windows in the shape of triangles. How much molding is needed to enclose five windows if a window measures 1 foot 7 inches on each side?

14. Joyce has 15 feet of lace for a border. How much lace will she have left if she puts a border on a triangular tablecloth that has sides of 4 feet 6 inches each?

4 ft 6 in.

4 ft 6 in.

4 ft 6 in.

AREA

The area of a **triangle** is given by the formula

$$A = \tfrac{1}{2}bh$$

where b stands for base and h stands for height.

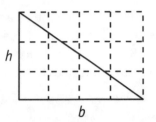

> **Note: 1.** When working with triangles, it is common to talk of **base** and **height** instead of *length* and *width*.
>
> **2.** As the drawing above shows, the area of a triangle is equal to $\frac{1}{2}$ the area of a rectangle with length b and width h.

The area formula is also used to find the area of triangles that do not contain a right angle.

The area of △ABC is equal to $\frac{1}{2}$ times the base (AC) times the height (BD). Notice that the height BD is not one of the sides of △ABC. The height BD is drawn in to show the distance between the vertex point B and the base AC.

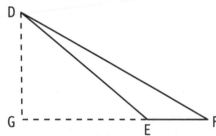

The area of △DEF is equal to $\frac{1}{2}$ times the base (EF) times the height (DG). Since the vertex point D does not lie directly over the base EF, extend the base EF to point G. The height is then the line DG.

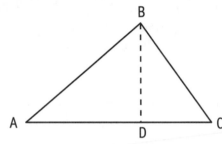

> **Note:** In each case above, the height that appears in the formula is a line **perpendicular** to the base or to its extension. Perpendicular lines meet at right angles.

<u>EXAMPLE 1</u> What is the area of the triangle at the right?

> **STEP 1** Identify the base (b) and the height (h).
>
> $b = 12$ in. and $h = 8$ in.
>
> **STEP 2** Substitute 12 for b and 8 for h in the area formula.
>
> $$A = \tfrac{1}{2}bh$$
>
> $$A = \tfrac{1}{2} \times 12 \times 8$$
>
> $$A = \mathbf{48}$$

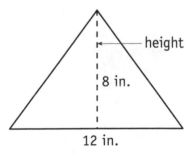

ANSWER: The area is 48 square inches.

<u>EXAMPLE 2</u> What is the area of ΔQRS?

STEP 1 Identify the base (*b*) and the height (*h*). Side RS is the base and the dotted line QT is the height.

$$b = 3 \text{ ft and } h = 5 \text{ ft}$$

STEP 2 Substitute 3 for *b* and 5 for *h* in the area formula.

$$A = \tfrac{1}{2}bh$$
$$A = \tfrac{1}{2} \times 3 \times 5$$
$$= \tfrac{15}{2} = 7\tfrac{1}{2}$$

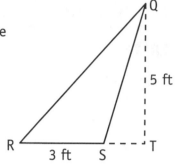

ANSWER: The area is $7\tfrac{1}{2}$ square feet.

 Use the formula $A = \tfrac{1}{2}bh$ to find the area of each triangle.

1. A = _____

8 in.

14 in.

2. A = _____

5 yd

9 yd

3. A = _____

5 cm

3 cm

4. A corner of a counter top is shaped like a right triangle. The two sides forming the right angle have lengths of 3 feet and $5\tfrac{1}{2}$ feet. What is the surface area of this corner?

3 ft

$5\tfrac{1}{2}$ ft

5. Jim cut a piece of plywood into the shape of a triangle. To the nearest square foot, find the area of the plywood if the base measures 8 feet and the height measures $6\tfrac{1}{4}$ feet.

6. A seamstress has a piece of cloth left over that is in the shape of an isosceles right triangle. How many square yards of cloth are in the piece if each of the two equal sides measures 3 yards? (**Hint:** The equal sides meet at a right angle.)

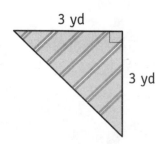

3 yd

3 yd

Working with Parallelograms and Trapezoids

PERIMETER

The perimeter of a **parallelogram** is given by the formula

$$P = 2s_1 + 2s_2$$

where s_1 stands for side 1
and s_2 stands for side 2.

Parallelogram

Note: The parallel sides of a parallelogram have equal lengths.

The perimeter of a **trapezoid** is found by adding its four sides.

$$P = s_1 + s_2 + s_3 + s_4$$

where s_1 stands for side 1,
s_2 stands for side 2,
s_3 stands for side 3, and
s_4 stands for side 4.

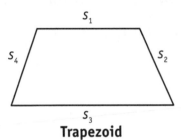

Trapezoid

EXAMPLE 1 What is the perimeter of the parallelogram at the right?

Substitute 7 for s_1 and 4 for s_2 in the formula.

$$P = 2s_1 + 2s_2$$
$$P = (2 \times 7) + (2 \times 4)$$
$$= 14 + 8 = \mathbf{22}$$

ANSWER: The perimeter is 22 feet or 7 yards 1 foot.

EXAMPLE 2 What is the perimeter of the trapezoid at the right?

To find the perimeter, add the four sides of the trapezoid.

```
   8
   6
   9
 + 5
 ‾‾‾‾
  28
```

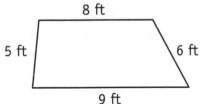

ANSWER: The perimeter is 28 feet or 9 yards 1 foot.

Find the perimeter of each figure.

1. $P =$ _____

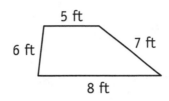

2 cm 4 mm
1 cm 7 mm

2. $P =$ _____

5 ft
6 ft
7 ft
8 ft

3. $P =$ _____

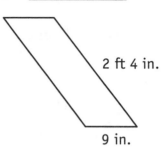

2 ft 4 in.
9 in.

4. $P =$ _____

3 ft 5 in.
3 ft 5 in.
3 ft 5 in.
3 ft 5 in.

5. $P =$ _____

5 km
4 km 300 m
6 km 800 m
5 km

6. $P =$ _____

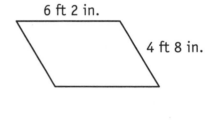

6 ft 2 in.
4 ft 8 in.

7. Dino's restaurant has a neon sign in the shape of a parallelogram. How much tubing does Dino need to enclose the sign in a neon edge if the sign has side lengths of 4 feet 5 inches and 3 feet 8 inches?

8. Joan's lot is in the shape of a trapezoid. What is the perimeter of the lot if the sides measure 79 m 85 cm, 68 m 50 cm, 73 m 60 cm, and 62 m 20 cm?

9. How much fencing will it take to enclose a garden that is shaped like a trapezoid if the sides measure 80 feet, 67 feet, 70 feet, and 61 feet?

10. A four-sided lot measures 235 yards 2 feet by 208 yards 1 foot by 225 yards by 211 yards. How many laps around the lot are equivalent to 2 miles? (1 mi = 1,760 yd)

11. How much molding is needed to go around a parallelogram-shaped window that measures 1 m 90 cm along one side and 1 m 6 cm along the other?

12. What is the distance around a rhombus, each side measuring 1 foot 4 inches? (**Remember:** A rhombus is a parallelogram that has four sides of the same length.)

AREA

The area of a **parallelogram** is given by the formula

$$A = bh$$

where b stands for base and h stands for height.

Parallelogram

> **Note: 1.** The **base** is one of the sides on the top or the bottom.
> **2.** The **height** is the direct distance between the base and its opposite side. As pictured, the height may be measured inside the parallelogram or outside by dropping a line from the top and extending the base if necessary.

EXAMPLE 1 What is the area of the parallelogram pictured at the right?

STEP 1 Identify the base (b) and the height (h).

$b = 13$ and $h = 7$

STEP 2 Substitute 13 for b and 7 for h in the area formula.

$$A = bh$$
$$A = 13 \times 7 = \mathbf{91}$$

7 in.

13 in.

ANSWER: The area is 91 square inches.

The area of a **trapezoid** is given by the formula

$$A = \tfrac{1}{2}(b_1 + b_2)h$$

where b_1 stands for base #1,
b_2 stands for base #2, and
h stands for height.

Trapezoid

> **Note: 1.** The **bases** are the pair of parallel sides.
> **2.** The **height** is the distance between the bases.

EXAMPLE 2 What is the area of the trapezoid pictured at the right?

STEP 1 Identify b_1 and b_2 and h.

$b_1 = 6$ and $b_2 = 9$ and $h = 5$

STEP 2 Substitute 6 for b_1, 9 for b_2, and 5 for h in the area formula.

$$A = \tfrac{1}{2} \times (6 + 9) \times 5$$

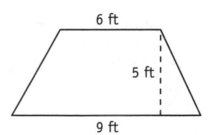

6 ft

5 ft

9 ft

STEP 3 Add the numbers inside the parentheses before multiplying.

$$A = \frac{1}{2} \times (15) \times 5$$

Next, remove the parentheses and multiply.

$$A = \frac{1}{2} \times 15 \times 5$$
$$= \frac{15 \times 5}{2}$$
$$= \frac{75}{2}$$
$$= 37\frac{1}{2}$$

ANSWER: The area is $37\frac{1}{2}$ square feet.

...

Find the area of each figure.

1. $A =$ _____

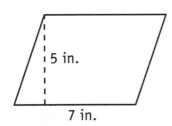

5 in.

7 in.

2. $A =$ _____

6 m

4 m

9 m

3. $A =$ _____

11 ft

10 ft

4. The basement of Stephen White's house is in the shape of a trapezoid. What is the area of the basement if the parallel sides have lengths of 38 feet and 34 feet, and the distance between these parallel sides is 22 feet?

5. A playground in the city park is shaped like a parallelogram. If the base measures 35 meters, and the height measures 27 meters, what is the area of this playground?

27 m

35 m

6. The Jones's Roofing Co. is going to shingle a section of roof that is in the shape of a trapezoid. The parallel sides measure 54 feet and 46 feet, and are 24 feet apart. If a bag of shingles covers 50 square feet, how many bags of shingles are needed to cover this roof?

Working with Circles

As you know, the curved figure to the right is a **circle**.

All points on the outside of a circle are at an equal distance from the center. This distance is called the **radius**. The symbol for radius is *r*.

The **diameter** is the distance across the circle. As you can see, the diameter passes through the center of the circle, and is equal to twice the radius. The symbol for diameter is *d*.

In symbols $d = 2r$ or $r = \frac{1}{2}d$.

The distance around the circle is called the **circumference**. The symbol for circumference is *C*.

CIRCUMFERENCE

The circumference of a circle is given by a formula

$$C = \pi d$$

where π has the approximate value of 3.14 or $\frac{22}{7}$.

To find the circumference, multiply the value of π by the diameter. Use the value $\frac{22}{7}$ for π when the diameter is given as a whole number or fraction. Use the value 3.14 for π when the diameter is given as a mixed decimal.

<u>EXAMPLE 1</u> Find the circumference of a circle that has a diameter of 14 inches.

Find the circumference by multiplying $\frac{22}{7}$ by 14.

$$\frac{22}{\underset{1}{7}} \times \overset{2}{14} = 44$$

ANSWER: 44 inches or 3 feet 8 inches

<u>EXAMPLE 2</u> Find the circumference of a circle that has a diameter of 9.1 feet.

Multiply 3.14 by 9.1 = **28.574**

ANSWER: 28.574 feet rounded to the nearest tenth is 28.6 feet.

If you use a calculator,

$$\boxed{C}\;\boxed{3}\;\boxed{.}\;\boxed{1}\;\boxed{4}\;\boxed{\times}\;\boxed{9}\;\boxed{.}\;\boxed{1}\;\boxed{=}\;\boxed{\;28.574\;} $$

 Use the formula $C = \pi d$ to find the circumference of each circle below. **Round your answers to the nearest tenth.**

1. C = _____

28 in.

2. C = _____

5.1 ft

3. C = _____

56 m

 In problems 4 through 6, find the diameter ($d = 2r$) as your first step.

4. $C =$ _____

2.5 ft

5. $C =$ _____

1.7 m

6. $C =$ _____

3 mi

 In problems 7 through 12, find the circumference. Notice that the words *distance across* are often used to indicate diameter, and the words *distance from center to edge* and *half the distance across* are used to indicate radius.

7. Green Reservoir is in the shape of a circle. If the distance across the reservoir is 4 miles, what is its circumference?

8. The distance between the center and the edge of a circular race track is 0.2 miles. What is the distance around the track?

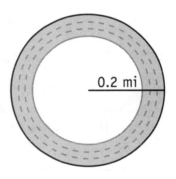

0.2 mi

9. To the nearest foot, what is the circumference of a circular swimming pool that has a radius of 6 feet?

10. Bill measured the widest distance across the top of a circular can to be 14 centimeters. What is the distance around the rim of the can?

11. Kelly plans to add a border around her circular tablecloth. Estimate the length of the piece she needs if the diameter of the tablecloth is 4.8 feet?

12. Half the distance across a circular garden measures 4.9 meters. Estimate how many meters of fencing material will be needed to enclose the garden.

4.9 m

AREA

The area of a **circle** is given by a formula

$$A = \pi r^2 \text{ which means } A = \pi \times r \times r.$$

To find the area of a circle, multiply the radius times itself and then multiply this product by π. As a first step, write the three numbers separated by times signs to see if cancellation is possible. Cancellation can greatly simplify multiplication.

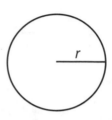

EXAMPLE 1 What is the area of a circle that has a radius of 7 inches?

STEP 1 Substitute $\frac{22}{7}$ for π, and 7 for *r* in the area formula.

$$A = \frac{22}{7} \times 7 \times 7$$

STEP 2 Cancel the first two 7's.

$$A = \frac{22}{7} \times \overset{1}{\cancel{7}} \times 7$$

STEP 3 Multiply 22 by 7.

$$A = 22 \times 7 = \mathbf{154}$$

ANSWER: 154 square inches

EXAMPLE 2 What is the area of a circle that has a radius of 2.1 feet?

STEP 1 Substitute 3.14 for π, and 2.1 for *r* in the area formula.

$$A = 3.14 \times 2.1 \times 2.1$$

STEP 2 Multiply 2.1 by 2.1.

$$2.1 \times 2.1 = 4.41$$

STEP 3 Multiply 3.14 by 4.41.

$$3.14 \times 4.41 = \mathbf{13.8474}$$

ANSWER: 13.8 square feet

 Use the formula $A = \pi r^2$ to find the area of each circle below. Round your answers to the nearest tenth.

1. $A =$ _____

14 in.

2. $A =$ _____

6.1 ft

3. $A =$ _____

21 km

 In problems 4 through 6, find the radius ($r = \frac{d}{2}$) as your first step in solving for the area.

4. $A =$ _____
(Use $\frac{22}{7}$.)

d = 14 m

5. $A =$ _____
(Use 3.14.)

d = 10 in.

6. $A =$ _____
(Use 3.14.)

d = 3.2 yd

Solve the area problems below.

7. A round glass mirror measures 1 foot from the center to the edge. How many square inches of glass are in the mirror?

8. To the nearest square foot, what is the area of a circular rose garden that measures 28 feet through the center?

9. A sprinkler waters in a circular pattern, spraying water out a distance of 3.1 meters. About how much lawn surface does the sprinkler water?

10. If six people share a large pizza that has a 14-inch diameter, how much pizza will each person get? Round your answer to the nearest square inch.

11. The circular flat roof of a water tower has a diameter of 29.5 feet. Estimate the area of this roof.

12. Radio station KNPA broadcasts its signal to all points within an 84 kilometer radius of the station. What is the area of the region served by this station?

Perimeter and Area: Applying Your Skills

Below are summarized the perimeter and area formulas for the six figures you've studied in this chapter. You should memorize how to find the perimeter and area of each of the first four figures. You should be familiar with how to use the formulas for the last two figures.

To Remember

Name	Example	Perimeter	Area
Square		$P = 4s$	$A = s^2$
Rectangle		$P = 2l + 2w$	$A = lw$
Triangle		$P = s_1 + s_2 + s_3$	$A = \frac{1}{2}bh$
Circle		$C = \pi d$ (circumference)	$A = \pi r^2$

Hint: 1. To remember **perimeter:**
For most figures, remember that perimeter is the sum of the sides. For a circle, memorize the circumference formula $C = \pi d$.

2. To remember **area:**
For most figures, remember to multiply one side by a second side that is perpendicular to the first. For a triangle, this often means multiplying by a height that may not be one of the sides and also remembering to multiply by $\frac{1}{2}$. The $\frac{1}{2}$ is easily remembered because a triangle looks very much like $\frac{1}{2}$ of a rectangle or parallelogram. For a **circle**, memorize the area formula $A = \pi r^2$.

3. Always remember, π is used only with circles.

To Be Familiar With

Name	Example	Perimeter	Area
Parallelogram		$P = 2s_1 + 2s_2$	$A = bh$
Trapezoid		$P = s_1 + s_2 + s_3 + s_4$	$A = \frac{1}{2}(b_1 + b_2)h$

 Solve the problems below. Round your answers to the nearest tenth. Use your calculator if you find it helpful.

1. P = _____

A = _____

5 ft

3 ft 7 in. 3 ft 3 ft 7 in.

5 ft

2. P = _____

A = _____

4 km

9 km 9 km

4 km

3. P = _____

A = _____

7 in.

$6\frac{1}{2}$ in. 6 in. $7\frac{1}{2}$ in.

14 in.

4. P = _____

A = _____

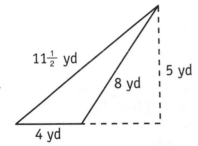

$11\frac{1}{2}$ yd 5 yd

8 yd

4 yd

5. C = _____

A = _____

10 cm

6. P = _____

A = _____

3 in. 6 in.

6 in. 3 in.

7. C = _____

A = _____

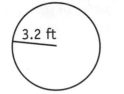

3.2 ft

8. P = _____

A = _____

2.4 cm 2.7 cm

2 cm

3.1 cm

9. P = _____

A = _____

3 ft 3 ft

3 ft 3 ft

 In the following problems, decide what shape you are working with and whether you are looking for perimeter or area. Then solve the problem. Round your answers to the nearest tenth. You may use your calculator.

10. Luis made a round oak table that he wants to cover with glass. How many square feet of glass are needed if the distance across the center of the table is 4 feet?

11. A swimming pool cover measures 5 meters long by 3 meters wide. At $5.25 per square meter, what is the cost of the cover?

12. After jogging 8 laps around a square field, Anne was curious about how far she had run. What was the total distance she had run if each side of the field measures 105 yards 2 feet?

13. In a carnival ride, the ponies walk in a circle with a 14-foot radius. How far does a child ride if the pony makes ten trips around before the child gets off?

14. The wall of Joan's living room measures 17 feet long and 8 feet high. To buy paint for the room, she needs to know the size of each wall. How many square feet is the living room wall she measured?

15. A rainbird sprinkler sends out water in a circular pattern. If the water reaches out a distance of 3.5 meters from the sprinkler, estimate how many square meters of lawn the sprinkler can water.

16. Central Park has three sides with lengths of 345 yards 2 feet, 464 yards 1 foot, and 500 yards. How far is it around this park?

17. At the base of Hyde Park is a circular water tower. If the diameter of this tower is 49 feet, what is the distance around it?

18. Marie owns a piece of property with four equal sides. What is the area of this property if each side measures 21 meters?

19. To increase the size of his house, Jim decided to enclose his garage. How many square feet can Jim add to the house if the garage measures 24 feet long by 25 feet wide?

20. Compute how much fence will it take to enclose a triangular garden space which has sides of 12 yards 2 feet, 14 yards 1 foot, and 13 yards 2 feet.

21. Estimate how much surface a wading pool has if the diameter of the circular pool measures 42 feet?

22. The roof on Meg's house consists of two rectangular sides. Each side is 45 feet long and 20 feet wide. What is the total area of roof on Meg's house?

23. Refer to the drawing below. Which of the following fractions results when the area of triangle ABC is divided by the area of the square?

 a. $\frac{2}{3}$

 b. $\frac{1}{4}$

 c. $\frac{1}{3}$

 d. $\frac{1}{2}$

 e. $\frac{3}{4}$

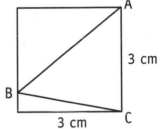

24. Refer to the drawing below. Choose the whole number that is closest to the result of dividing the area of the square by the area of the circle.

 a. 4
 b. 5
 c. 6
 d. 7
 e. 8

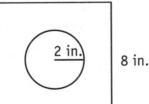

25. George's backyard is in the shape of a trapezoid. What is the distance around the yard if the sides measure 14 m 50 cm, 12 m 80 cm, 15 m 75 cm, and 21 m 25 cm?

Solving Two-Step Area Problems: Part I

Many area problems involve figures that are a combination of common geometrical shapes. We call these problems **two-step area problems.**

To solve a two-step area problem, divide the figure into shapes you are familiar with and then solve for each shape separately.

For example, you may be asked to find the area of the room pictured to the right. To find this area, follow these steps.

STEP 1 Divide the room into two rectangles.

STEP 2 Find the unknown side of one rectangle.

STEP 3 Find the area of each rectangle.

STEP 4 Add the areas of the two rectangles. The example below will show you how to do this.

EXAMPLE What is the area of the room pictured at the right?

STEP 1 Divide the room into two rectangles. Label the rectangles I and II. Label the unmeasured long side of rectangle I as l for length. You need to know the value of l before you can find the area of rectangle I.

STEP 2 To find l subtract 8 from 24.

$$l = 24 - 8$$
$$l = 16$$

STEP 3 Find the area of rectangle I and the area of rectangle II.

Area of I = lw = 16 × 10
= 160 sq ft

Area of II = lw = 18 × 8
= 144 sq ft

STEP 4 Add the areas of rectangles I and II.

160 sq ft
+144 sq ft
304 sq ft

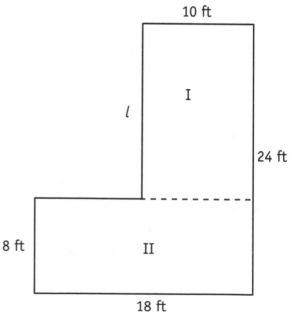

ANSWER: The area of the room is 304 sq ft.

What is the area of each room? You may use your calculator.

1. $A =$ _____

2. $A =$ _____

3. $A =$ _____

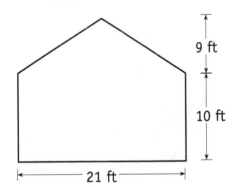

(**Hint:** Problem 3 involves a rectangle and a triangle.)

4. $A =$ _____

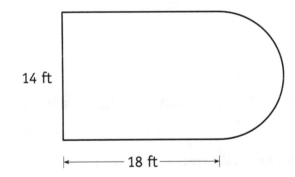

(**Hint:** Problem 4 involves a rectangle and half of a circle. Do you see a clue for the circle's diameter?)

Solving Two-Step Area Problems: Part II

Many area problems require that you subtract a smaller area from a larger area.

EXAMPLE How many square feet of wallpaper will be needed to cover the wall pictured at right?

To find the area of the wall, first find the area of the wall including the window and then subtract the area of the window.

8 ft 3 ft 5 ft

14 ft

STEP 1 Find the area of the wall including window.

Area of the wall including window = 14 × 18
 = 112 sq ft

STEP 2 Find the area of the window.

Area of window = 5 × 3
 = 15 sq ft

STEP 3 Subtract to find the area of the wall.

 112 sq ft
 − 15 sq ft
 97 sq ft

ANSWER: The amount of wallpaper needed is 97 square feet.

 Solve the following problems. You may use your calculator.

1. How many square feet of wallpaper will be needed to cover the wall pictured below?

3 ft

7 ft

8 ft

11 ft

2. How many square feet of tile are needed to cover the patio shown below?

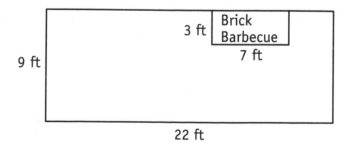

3 ft Brick Barbecue

7 ft

9 ft

22 ft

3. Lois's house sits in the middle of her lot. The lot is shaped like a trapezoid. Except for the house and driveway, her lot is covered with grass. How many square feet of grass are on the lot?

4. The Parks Department wants to put tile around the outside of the circular wading pool. Estimate the number of square feet of tile that will be needed.

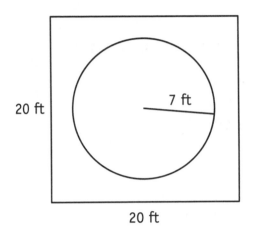

5. Sam, a machinist, wants to cut a circular disc from a square plate of aluminum. From the drawing below, how many square centimeters of surface must be cut away?

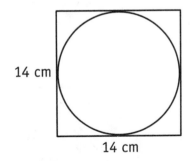

6. Two triangular planters sit in the rose garden. Excluding the planters, how many square feet of garden area remain?

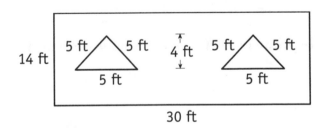

Plane Figures Review

This review covers the material you have just studied.

Add, subtract, multiply, or divide.

1. 7 yd 2 ft
 3 yd 1 ft
 + 8 yd 2 ft

2. 9 cm 7 mm
 6 cm 8 mm
 + 3 cm 4 mm

3. 7 ft 6 in.
 − 3 ft 9 in.

4. 35 m 87 cm
 − 14 m 93 cm

5. 1 m 75 cm
 × 7

6. 6)‾14 ft 6 in.

7. 4)‾5 m 12 cm

8. LaToya measured 19 inches on the day she was born. On her fourth birthday, she was 3 feet 5 inches tall. By how much did LaToya grow in her first four years?

9. Jesse, a carpenter, wants to cut a 17 foot long board into three equal pieces. To the nearest inch, how long will each piece be?

In problems 10 through 15, name each of the figures shown.

10. _____

11. _____

12. _____

13. _____

14. _____

15. _____

16. Match each formula on the left with the area it represents.

_____ **1.** bh **a.** area of a square

_____ **2.** $\frac{1}{2}(b_1 + b_2)h$ **b.** area of a circle

_____ **3.** s^2 **c.** area of a parallelogram

_____ **4.** $\frac{1}{2}bh$ **d.** area of a rectangle

_____ **5.** πr^2 **e.** area of a trapezoid

_____ **6.** lw **f.** area of a triangle

 In problems 17 through 22, find the perimeter and area of each figure. You may use your calculator.

17. $P =$ _____

$A =$ _____

7 ft $8\frac{1}{2}$ ft

11 ft

18. $P =$ _____

$A =$ _____

5 in.

5 in.

19. $C =$ _____

$A =$ _____

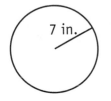
7 in.

20. $P =$ _____

$A =$ _____

10 ft $11\frac{1}{2}$ ft

8 ft

$14\frac{1}{4}$ ft

21. $P =$ _____

$A =$ _____

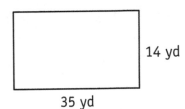
14 yd

35 yd

22. $P =$ _____

$A =$ _____

15 mi

9 mi 10 mi 9.5 mi

7.6 mi

 23. Melvin wants to put a strip of tile around his circular swimming pool. What is the distance around the pool if the radius of the pool is 7 feet?

24. Find the area of Sally's lot which is in the shape of a square, each side measuring 35 meters?

25. Find the floor area of the recreation room pictured below.

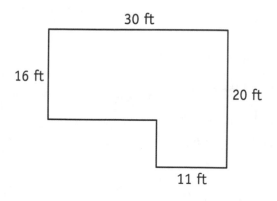

26. How many square feet of paneling will be needed to cover the wall drawn below?

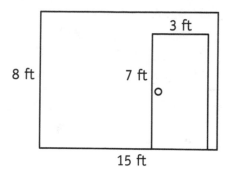

Plane Figures Review Chart

Circle the number of any problem that you missed and review the appropriate pages. A passing score is 22 correct answers. If you miss more than four questions, you should review this chapter.

PROBLEM NUMBERS	SKILL AREA	PRACTICE PAGES
1, 2	addition: U.S. and metric units	82–83
3, 4	subtraction: U.S. and metric units	82–83
5	multiplication: metric units	84–85
6, 7	division: U.S. and metric units	84–85
8, 9	length units: applying skills	86–87
10, 11, 12, 13, 14, 15	recognizing shapes	90–91, 110
16	formulas	114
17	parallelogram: perimeter + area	106–109
18	square: perimeter + area	94–97
19	circle: circumference + area	110–113
20	triangle: perimeter + area	102–105
21	rectangle: perimeter + area	98–101
22	trapezoid: perimeter + area	106–109
23, 24	perimeter + area: applying skills	114–117
25, 26	two-step area problem	118–121

SOLID FIGURES

Solid Figures Skills Inventory

This inventory will let you know what you need to work on in the Solid Figures section of *Geometry*. Do all of the following problems that you can. There is no time limit. On this inventory, you can use $\frac{22}{7}$ for π. If you use a calculator, use 3.14 for π.

1. Match each formula with the volume it represents.

 _____ **1.** $\frac{1}{3}\pi r^2 h$ **a.** volume of a rectangular solid

 _____ **2.** s^3 **b.** volume of a cylinder

 _____ **3.** lwh **c.** volume of a cone

 _____ **4.** $\pi r^2 h$ **d.** volume of a cube

In problems 2 through 5, find the volume of each figure.

2. $V =$ _____

8 in.
8 in.
8 in.

3. $V =$ _____

5 yd
17 yd
6 yd

4. $V =$ _____

4 ft
$r = 1$ ft

5. $V =$ _____

21 cm
$r = 7$ cm

6. What is the volume of a room measuring 10 feet on each side?

7. Find the volume of a storage shed that measures 50 meters long, 35 meters wide, and 6 meters high.

8. How many cubic feet of water will a cylindrical storage tank hold if it is 15 feet high and has a radius of 14 feet?

9. A pile of sand is in the shape of a cone. Estimate the volume of sand that is in a pile measuring 4 yards high and having a diameter of 10 yards.

In problems 10 and 11, find the volume of each figure.

10. $V =$ _____

11. $V =$ _____

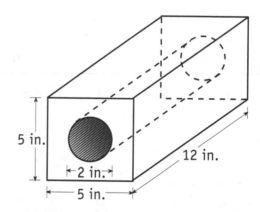

Recognizing Common Solid Figures

Solid figures take up space and have volume. Things you can see, touch, and hold are solid figures. There are four shapes of solid figures that occur so frequently that they are of special interest in the study of geometry. Below are listed the names and descriptions of these common solid figures.

Name	Example	Description
Cube		Each side of a **cube** has an equal length, and each pair of sides forms a right angle. Thus, each **face** (flat surface) of a cube is a square.
Rectangular Solid	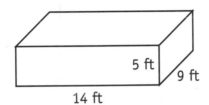	Each face of a **rectangular solid** is either a rectangle or a square. At every corner, each pair of sides forms a right angle. A rectangular solid is also called a **rectangular prism.**
Cylinder	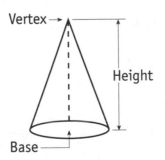	A **cylinder** has the shape of a common tin can. The top and bottom surfaces are circles that are parallel to each other. The distance between the top and bottom is called the **height** of the cylinder.
Cone		A **cone** has one circular surface called the **base.** The **vertex** of a cone is a point that lies directly above the center of the base. The distance between the vertex and the center of the base is called the **height** of the cone.

Label each figure below with its solid figure name: cube, rectangular solid, cylinder, or cone.

1. _____

City Water Storage

2. _____

Sugar

3. _____

Cereal

4. _____

Ice Cream Cone

5. _____

Grain Silo

6. _____

Dictionary

7. _____

8. _____

9. _____

Pile of Sand

10. _____

11. _____

Party Hat

12. _____

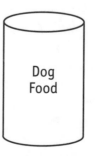

Dog Food

Volume

Volume is a measure of the space *taken up* by a solid figure (or object). A large object has more volume than a small object. For example, the volume of a brick is greater than the volume of a marble.

Volume can also refer to the space *enclosed* by the surface of a solid figure. For example, the volume of a room is the space enclosed by the floor, walls, and ceiling of the room.

Questions on volume may refer to either use of the word volume. In either case, the symbol for volume is *V.*

MEASURING VOLUME

To measure volume, you can use a **volume unit** in the shape of a cube.

For example, you can use this cube as your volume unit. The volume of solid objects is measured in cubic units.

Volume can be found by first dividing an object into volume units and then counting these units.

EXAMPLE What is the volume of the rectangular solid at the right?

The rectangular solid is divided into volume units. To find the volume, count these units. Because you can't directly see each volume unit, follow these steps:

STEP 1 Count the number of volume units in one layer of the figure.
There are 12 volume units in the front layer.

Multiply the volume units in the length times the volume units in the height.

Area of surface = 3 × 4 = 12

STEP 2 Multiply the number of volume units in the front layer (12) by the number of layers (3) in the figure.

12 × 3 = **36**

ANSWER: *V* = 36 volume units

Find the volume of each figure below. Each figure is divided into volume units.

1. $V =$ _____ volume units

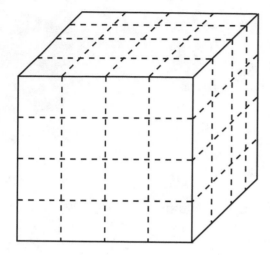

2. $V =$ _____ volume units

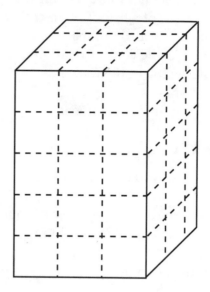

3. $V =$ _____ volume units

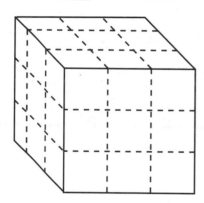

4. $V =$ _____ volume units

5. $V =$ _____ volume units

6. $V =$ _____ volume units

Working with Cubes

VOLUME UNITS

The volume of common geometrical figures is most often found by using a **volume formula.** Using a volume formula is easier than trying to divide a figure into volume units. To use a volume formula, replace letters with numbers and then multiply or divide as indicated. Use the correct volume unit with each answer.

Common volume units include the following:

U.S. Units	Abbreviation
cubic inch	cu in.
cubic foot	cu ft
cubic yard	cu yd
cubic mile	cu mi

Metric Units	Abbreviation
cubic millimeter	mm^3
cubic centimeter	cm^3
cubic meter	m^3
cubic kilometer	km^3

VOLUME OF A CUBE

The volume of a cube is given by the formula

$$V = s^3$$

where s stands for *side* and s^3 means $s \times s \times s$.

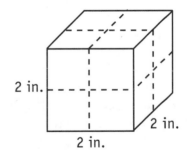

To find the volume of a cube, multiply the value of a side by itself and then multiply this product by the value of the side once more. The volume of the cube at the right is $2 \times 2 \times 2$ or 8 cu in.

EXAMPLE What is the volume of the cube pictured below?

STEP 1 Replace each s with 4 in the volume formula.

$$V = s^3$$
$$V = 4^3$$

STEP 2 Multiply and solve.

$$V = 4 \times 4 \times 4$$
$$V = \mathbf{64}$$

Since each side is measured in feet, the volume units are written as cubic feet.

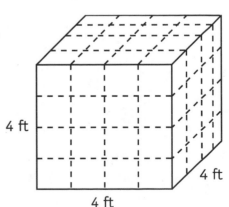

ANSWER: The volume is 64 cubic feet.

Use the formula $V = s^3$ to find the volume of each cube below.

1. V = _____

7 in.

2. V = _____

3 cm

3. V = _____

8 yd

4. A contractor is hired to put fill dirt in a hole that is in the shape of a cube. If the hole measures 18 yards on an edge, how much dirt will the contractor need?

5. Compute the volume of a storage shed that has a square floor 3 meters on each side and is 3 meters high.

6. A water storage tank in the shape of a cube measures 4.1 meters along each edge. If one cubic meter of water weighs one metric ton, how many metric tons of water can the tank hold?

7. The Deluxe Calculator Co. packs calculators in cubical shipping boxes that are 2 feet on each side. How many shipping boxes can be put in a moving van that has a packing volume of 896 cubic feet?

8. Simon's Toy Co. makes plastic blocks that are in the shape of cubes 3 inches on each edge. How many blocks can be put in a cube-shaped mailing box that measures 15 inches on each edge?

Working with Rectangular Solids

VOLUME

The volume of a rectangular solid is given by the formula

$$V = lwh$$

where l stands for length,
 w stands for width, and
 h stands for height.

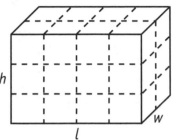

To find the volume of a rectangular solid, multiply the length by the width, and then multiply this product by the height.

EXAMPLE What is the volume of the rectangular solid at right?

 STEP 1 Substitute 3 for l, 2 for w, and 4 for h in the volume formula.

$$V = lwh$$
$$V = 3 \times 2 \times 4$$

 STEP 2 Multiply and solve.

$$V = 24$$

 Write the volume as cubic centimeters.

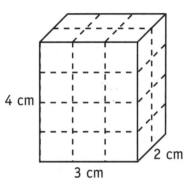

ANSWER: The volume is 24 cm³.

 Use the formula $V = lwh$ to find the volume of each rectangular solid.

1. $V =$ _____

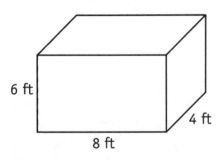

6 ft 4 ft 8 ft

2. $V =$ _____

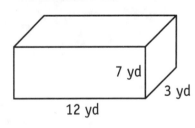

7 yd 3 yd 12 yd

3. $V =$ _____

8 cm 10 cm 14 cm

4. V = _____

10 in.

5 in.

7 in.

5. V = _____

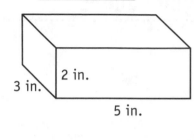

2 in.

3 in.

5 in.

6. V = _____

6 m

3 m

2 m

7. How many cubic feet of water are in a waterbed that is 6 feet long, 5 feet wide, and $\frac{1}{2}$ foot high?

5 ft

$\frac{1}{2}$ ft

6 ft

8. Find the volume of a storage chest that measures 3 feet long, 2 feet wide, and 2 feet high.

2 ft

3 ft

2 ft

9. What is the volume of space in a bedroom that measures 4 meters long, 3 meters wide, and $2\frac{1}{2}$ meters high?

10. A freezer has dimensions of 4 feet, 2 feet, and $2\frac{1}{2}$ feet. About how many cubic feet are in the freezer?

4 ft

$2\frac{1}{2}$ ft

2 ft

Working with Cylinders

VOLUME

The volume of a cylinder is given by the formula

$$V = \pi r^2 h$$

where r stand for radius
and h stands for height.

This formula means the volume is equal to π times the radius squared times the height. You can use either $\frac{22}{7}$ or 3.14 for π. Your answer may be slightly different depending on which number you use.

The top and bottom of a cylinder are equal-size circles.

EXAMPLE What is the volume of the cylinder at the right?

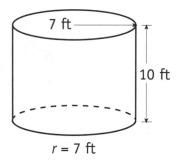

STEP 1 Substitute the values in the problem.

$$V = \pi r^2 h$$
$$V = \frac{22}{7} \times 7^2 \times 10$$
(or $V = 3.14 \times 7^2 \times 10$)

STEP 2 Find the radius squared.

$$V = \frac{22}{7} \times 49 \times 10$$
(or $V = 3.14 \times 49 \times 10$)

STEP 3 Cancel, if possible. Multiply and solve.

$$V = \frac{22}{_1 \cancel{7}} \times \cancel{49}^{\,7} \times 10$$
$$V = \mathbf{1,540}$$
(or $V = \mathbf{1538.6}$)

Write the volume unit as cubic feet.

ANSWER: The volume is 1,540 (or 1538.6) cubic feet.

 Use the formula $V = \pi r^2 h$ to find the volume of the cylinders below. Use 3.14 for π if you use your calculator or $\frac{22}{7}$ if you use pencil and paper.

1. $V =$ _____

$r = 1$ m

2. $V =$ _____

$r = 5$ in.

3. $V =$ _____

8 ft
$r = 3.5$ ft

Working with Cones

VOLUME

The volume of a cone is given by the formula

$$V = \tfrac{1}{3}\pi r^2 h$$

where r stands for radius
and h stands for height.

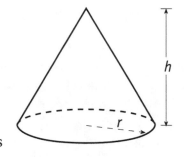

The volume of a cone is equal to $\tfrac{1}{3}$ the volume of a cylinder that has the same base area and height.

EXAMPLE What is the volume of the cone at the right?

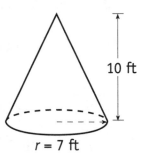

10 ft

$r = 7$ ft

STEP 1 Substitute the values in the formula.

$$V = \tfrac{1}{3}\pi r^2 h$$

$$V = \tfrac{1}{3} \times \tfrac{22}{7} \times 7^2 \times 10$$

(or $V = \tfrac{1}{3} \times 3.14 \times 7^2 \times 10$)

STEP 2 Find the radius squared.

$$V = \tfrac{1}{3} \times \tfrac{22}{7} \times 49 \times 10$$

(or $V = \tfrac{1}{3} \times 3.14 \times 49 \times 10$)

STEP 3 Cancel, if possible. Multiply and solve.

$$V = \tfrac{1}{3} \times \tfrac{22}{\underset{1}{7}} \times \overset{7}{49} \times 10 \qquad (\text{or } V = \tfrac{1}{3} \times 3.14 \times 49 \times 10$$

$$V = \frac{22 \times 7 \times 10}{3} \qquad\qquad V = \frac{3.14 \times 49 \times 10}{3}$$

$$V = \frac{1{,}540}{3} \qquad\qquad V = \frac{1538.6}{3}$$

$$V = 513\tfrac{1}{3} \qquad\qquad V = \textbf{512.9})$$

ANSWER: The volume is $513\tfrac{1}{3}$ (or 512.9) cubic feet.

Use the formula $V = \tfrac{1}{3}\pi r^2 h$ to find the volume of the cones below. Use **3.14 for π if you use your calculator or $\tfrac{22}{7}$ if you use pencil and paper.**

1. $V =$ _____

20 ft

$r = 14$ ft

2. $V =$ _____

2 ft

$r = 1$ ft

3. $V =$ _____

3 cm

$r = 2.5$ cm

Volume: Applying Your Skills

Learn the figure names and volume formulas as you practice your skills with word problems.

Name	Example	Volume Formula
Cube		$V = s^3$
Rectangular Solid		$V = lwh$
Cylinder		$V = \pi r^2 h$
Cone		$V = \frac{1}{3}\pi r^2 h$

 For the problems below, decide which formula is needed and solve. You may use your calculator.

1. How many cubic feet of concrete will be needed for a driveway section that is 10 feet long, 8 feet wide, and $\frac{1}{2}$ foot thick?

2. A cylindrical water tower is half full of water. How many cubic feet of water are in the tower if it measures 20 feet high and has a radius of 7 feet?

3. A pile of sand is shaped like a cone. If the height of the pile is 7 meters and the radius of the pile is 6 meters, how many cubic meters of sand are in the pile?

4. How many cubic yards of sand will it take to fill a child's play area that measures 10 yards long, 8 yards wide, and $\frac{1}{3}$ yard deep?

5. What is the volume of a storage area measuring 5 feet on each side?

6. A pile of gravel is in the shape of a cone. If the pile is 8 feet high, and the diameter of the pile is 10 feet, how many cubic feet of gravel are in the pile? (**Hint:** If the diameter is 10, what is the radius?)

7. A cylindrical oil drum is 2.1 meters high and has a radius of 0.9 meter. Estimate the number of cubic meters of oil the drum can hold.

8. How many cubic yards of topsoil are needed to put in a lawn if the lot measures 50 feet by 80 feet and the topsoil is to be $\frac{1}{3}$ foot thick? (**Hint:** As your first step, change all the feet to yards.)

9. What is the volume of a sugar cube that measures 1 cm on each side?

10. A glass is 7 inches high and has a diameter of 3 inches. How many cubic inches of water can the glass hold when filled to the top?

Solving Two-Step Volume Problems: Part I

Many volume problems involve figures that combine common geometrical shapes. We call these problems **two-step volume problems.**

To solve a two-step volume problem, divide the figure into shapes with which you are familiar and then find the volume of each part separately.

For example, to find the volume of the figure at the right, follow these steps.

STEP 1 Divide the figure into two rectangular solids.

STEP 2 Find the volume of each rectangular solid.

STEP 3 Add the volumes of the two rectangular solids found in Step 2.

The example below illustrates this method.

<u>EXAMPLE</u> What is the volume of the figure pictured at the right?

STEP 1 Divide the figure into two rectangular solids. Label these I and II.

STEP 2 Find the volume of each rectangular solid. Volume of figure I = lwh

$$V = 4 \times 2 \times 3$$
$$V = 24 \text{ cubic feet}$$

Volume of figure II = lwh
$$V = 9 \times 2 \times 7$$
$$V = 126 \text{ cubic feet}$$

STEP 3 Add the volumes of I and II.

24 cubic feet
+126 cubic feet
150 cubic feet

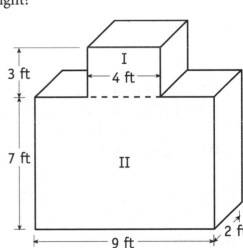

ANSWER: The volume of the figure is 150 cubic feet.

 Solve the following problems. You may use your calculator.

1. What is the volume of the figure drawn below?

2. How many cubic inches of concrete are needed for the section of curb shown below?

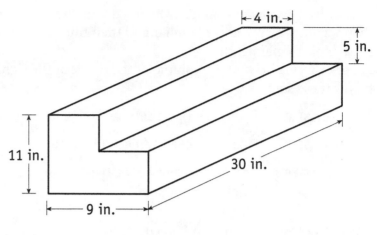

3. What is the volume of the mailbox shown below? (**Hint:** The mailbox can be thought of as a half-cylinder plus a rectangular solid.)

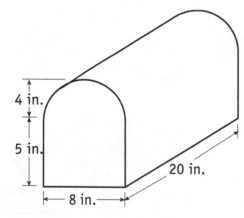

4. What is the volume of the figure below? (**Hint:** The figure can be separated into a cylinder and a cone.)

Solving Two-Step Volume Problems: Part II

Many volume problems are solved by subtracting a smaller volume from a larger volume.

EXAMPLE How many cubic inches of metal are contained in the rectangular bar at the right?

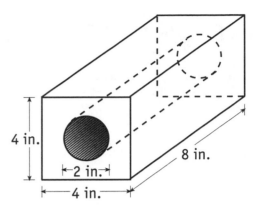

STEP 1 Find the volume of the bar as if it were solid.

Volume of solid bar = *lwh*
$$= 8 \times 4 \times 4$$
$$= 128 \text{ cubic inches}$$

STEP 2 Find the volume of space taken by the hole drilled through the bar.

Volume of hole in bar = $\pi r^2 h$
$$= \tfrac{22}{7} \times 1 \times 1 \times 8 \qquad (\text{or} = 3.14 \times 1 \times 1 \times 8 = 25.1)$$
$$= \tfrac{176}{7}$$
$$= 25\tfrac{1}{7} \text{ cubic inches}$$

STEP 3 Subtract to find the volume of metal in the bar. Write 128 as $127\tfrac{7}{7}$.

$$128\tfrac{7}{7} \text{ cubic inches} \qquad (\text{or} \qquad 128.0 \text{ cubic inches}$$
$$- \ 25\tfrac{1}{7} \text{ cubic inches} \qquad\qquad - \ 25.1 \text{ cubic inches}$$
$$\overline{\ 102\tfrac{6}{7} \text{ cubic inches}} \qquad\qquad \overline{102.9 \text{ cubic inches})}$$

ANSWER: The amount of metal left in the bar is $102\tfrac{6}{7}$ (or 102.9) cubic inches.

 Solve the following problems. You can use your calculator.

1. How many cubic inches of metal are contained in the hollow rectangular bar below?

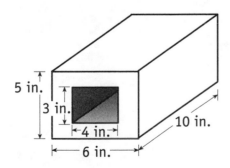

2. The concrete steps below contain how many cubic feet of concrete?

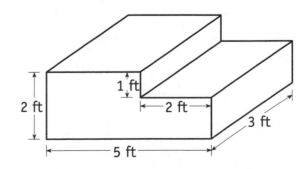

3. What is the volume of the slotted beam below?

4. How many cubic inches of metal are contained in the rectangular box below?

5. Find the volume of metal in the tube shown below.

6. Estimate the number of cubic inches of metal that are contained in the cylinder drawn below.

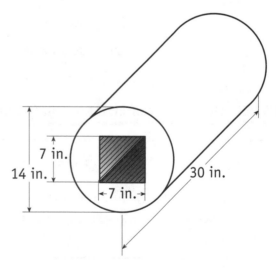

SOLID FIGURES REVIEW

This review covers the material you have just studied.

In problems 1 through 3, name each figure shown.

1. _____

2. _____

3. _____

4. Match each formula on the left with the volume it represents.

_____ **1.** lwh **a.** volume of a cylinder

_____ **2.** $\pi r^2 h$ **b.** volume of a cube

_____ **3.** $\frac{1}{3}\pi r^2 h$ **c.** volume of a rectangular solid

_____ **4.** s^3 **d.** volume of a cone

 In problems 5 through 8, find the volume of each figure shown. You may use $\frac{22}{7}$ or 3.14 for π.

5. $V =$ _____

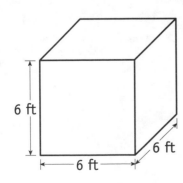

6 ft
6 ft
6 ft

6. $V =$ _____

5 ft
3 in.
11 in.

7. V = _____

20 yd

r = 7 yd

8. V = _____

39 in.

r = 21 in.

9. How many cubic yards of dirt will it take to fill a hole that is in the shape of a cube 3 yards on each side?

10. Decide how many cubic feet of space are in a moving van that is shaped like a rectangular solid and has the following dimensions: length = 20 feet, width = 7 feet, and height = 8 feet.

7 ft

20 ft

8 ft

11. A cylindrical grain silo is 100 feet high and has a radius of 14 feet. How many cubic feet of grain can be kept in the silo?

r = 14 ft

100 ft

12. A pile of gravel is in the shape of a cone. Estimate the number of cubic yards of gravel that are contained in a pile 10 yards high and 14 yards in diameter.

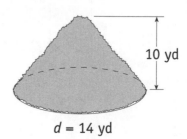

10 yd

d = 14 yd

In problems 13 through 14, find the volume of each figure. You may use $\frac{22}{7}$ or 3.14 for π.

13. $V =$ _____

14. $V =$ _____

Solid Figures Review Chart

Circle the number of any problem that you missed and review the appropriate pages. A passing score is 12 correct answers. If you miss more than three questions, you should review this chapter.

PROBLEM NUMBERS	SKILL AREA	PRACTICE PAGES
1, 2, 3	recognizing solid figures	127–128
4	formulas	137
5	cube: volume	131–132
6	rectangular solid: volume	133–134
7	cylinder: volume	135
8	cone: volume	136
9, 10, 11, 12	volume: applying skills	137–138
13, 14	two-step volume problem	139–142

Geometry Posttest A

This posttest gives you a chance to check your skills in applying geometry to the type of questions asked on the GED and other tests. Take your time and work each problem carefully. When you finish, check your answers and review any topics on which you need more work.

1. How many degrees are in one-fourth of a circle?

2. As shown at the right, a pie is cut into eight equal pieces. At what acute angle is each piece cut?

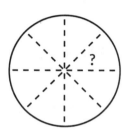

Problems 3 and 4 refer to the drawing at the right.

3. If Harlow Road intersects I-5 at a right angle, at what acute angle does Harlow Road intersect I-40?

4. What is the measure of the obtuse angle that I-5 makes with I-40?

Problems 5 and 6 refer to the drawing at the right.

5. At what acute angle does the Amtrak line cross Madison Street?

(**Note:** Johnson St. is parallel to Madison St.)

6. What is the measure of the obtuse angle that the Amtrak line makes with Madison Street near the Apex Warehouse?

7. What is the roof angle (angle b) in the drawing at the right?

8. In triangle ABC, $\angle A = 34°$ and $\angle B = 47°$. What is the measure of $\angle C$?

9. A surveyor wants to find the distance across Yellowstone River at Forest Camp. Use the drawing at the right that she made to determine this distance.

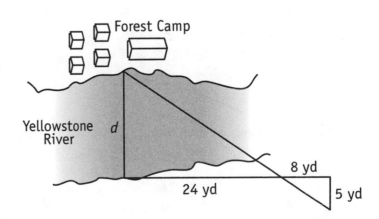

10. At 2:00 P.M., a 6-foot tall man cast an 8-foot shadow. If, at the same time, a nearby tree cast a 100-foot shadow, how high is the tree?

11. Refer to the map at the right and determine the direct distance (d) between Susanville and Demont.

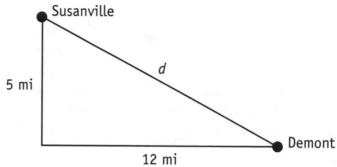

12. One side of a right triangle measures 6 meters. If the hypotenuse measures 10 meters, what is the length of the third side?

13. Mario bought 100 feet of fencing material to enclose his rectangular garden. Subtracting 3 feet for a gate, how much material will he have left over if his garden measures 27 feet 9 inches by 19 feet 5 inches?

14. How many square yards of carpet are needed to carpet a rectangular room that measures 4 yards by 5 yards?

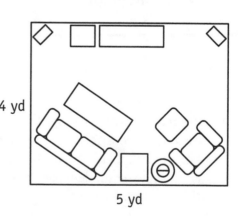

15. What is the area of the triangle drawn at the right?

16. Jason plans to place a piece of oak trim along the edge of a circular table he built. If the table is 48 inches across, what is the length of trim he needs? Express your answer to the nearest inch. (**Remember:** $C = \pi d$)

Jason's Table

48 inches

17. Use the area formula for a trapezoid, $A = \frac{1}{2}(b_1 + b_2)h$, to find the area of the figure below.

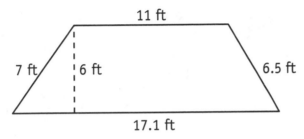

11 ft

7 ft 6 ft 6.5 ft

17.1 ft

18. Refer to the drawing at the right. What is the ratio of the area of the circle to the area of the square? Use $\frac{22}{7}$ for π.

$$\left(\text{ratio} = \frac{\text{area of circle}}{\text{area of square}} \right)$$

14 m

19. Bill's house sits in the middle of his lot which is in the shape of a rectangle. Except for the driveway and the house, the lot is planted in grass. How many square feet of the lot are planted in grass?

45 ft

65 ft House 30 ft

18 ft Driveway

8 ft

100 ft

20. What is the storage volume of a cabinet that has inside dimensions of 7 feet high by 4 feet wide by $2\frac{1}{2}$ feet deep?

7 ft

$2\frac{1}{2}$ ft

4 ft

21. The large block of clay at the right is to be cut up into a number of small cubes each measuring 2 inches on a side. How many of these smaller cubes can be cut from the large block? (**Hint:** How many cubic inches are in a 2-inch cube?)

6 in.

12 in.

8 in.

22. Use the volume formula for a cone, $V = \frac{1}{3}\pi r^2 h$, to find the volume of the pile of gravel shown at the right. Use 3.14 for π.

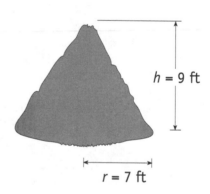

$h = 9$ ft

$r = 7$ ft

23. A contractor is pouring 12-foot long sections of curb as shown at the right. How many 12-foot sections of curb can he pour for each 108 cubic feet of concrete he mixes?

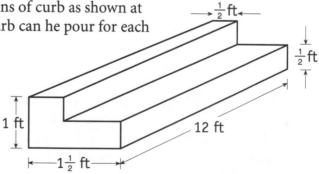

$\frac{1}{2}$ ft

$\frac{1}{2}$ ft

1 ft

12 ft

$1\frac{1}{2}$ ft

 24. A machinist "turns" a square rod and makes a cylindrical bar. As shown at the right, the diameter of the finished cylindrical bar equals the width of the uncut rod. For a 14 inch length of rod, estimate how much metal must be removed to make this bar round.

14 in.

2 in.

Geometry Posttest A Prescriptions

Circle the number of any problem that you miss. A passing score is 20 correct answers. If you passed the test, go on to Using Number Power. If you did not pass the test, review the chapters in this book or refer to these practice pages in other materials from Contemporary Books.

PROBLEM NUMBERS	PRESCRIPTION MATERIALS	PRACTICE PAGES
1, 2, 3, 4, 5, 6, 7	angles	14–33
	Math Exercises: Geometry	3–9
	Real Numbers: Geometry Basics	5–10, 64–68
8, 9, 10, 11, 12	triangles	42–69
	Math Exercises: Geometry	10–19
	Real Numbers: Geometry Basics	21–23
13	perimeter	88–89, 94–117
	Math Exercises: Geometry	20
	Real Numbers: Geometry Basics	26–32, 42–45
14, 15, 16, 17, 18, 19	area	92–121
	Math Exercises: Geometry	22–25
	Real Numbers: Geometry Basics	33–40, 42–49
21, 22, 23, 24	volume	127–142
	Math Exercises: Geometry	26–27
	Real Numbers: Geometry Basics	50–58

Geometry Posttest B

..

This posttest gives you a chance to check your geometry skills in a multiple-choice format as used in many standardized tests. Take your time and work each problem carefully before you choose your answer.

1. How many degrees are in angle A, the angle formed by the shaded part of the circle?

 a. 8°
 b. 12.5°
 c. 37.5°
 d. 45°
 e. 62.5°

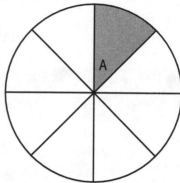

All sections of the circle are the same size.

2. Which angle could you add to a 90° angle to form an *obtuse angle*?

 a. **b.** **c.** **d.** **e.**

3. ∠A is the complement of ∠B. If ∠A measures 38°, what is the measure of ∠B?

 a. 19°
 b. 38°
 c. 52°
 d. 142°
 e. 322°

4. Fifth Avenue intersects Walnut Street at an acute angle of 29°. At what obtuse angle do these streets intersect?

 a. 29°
 b. 61°
 c. 90°
 d. 151°
 e. 331°

5. What is the sum of the measures of ∠A + ∠B?

 a. 70°
 b. 90°
 c. 110°
 d. 140°
 e. 220°

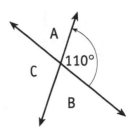

6. What is the *least number* of acute angles that a triangle can have?

 a. 0
 b. 1
 c. 2
 d. 3
 e. 4

7. What is the measure of ∠B in triangle ABC?

 a. 24°
 b. 34°
 c. 56°
 d. 90°
 e. 114°

∠A + ∠C = 124°

8. A ladder is leaning against the side of Jason's house. The ladder makes an angle of 68° with the ground. What angle does the ladder make with the house?

 a. 22°
 b. 30°
 c. 38°
 d. 45°
 e. 112°

9. The front side of a small camping tent is in the shape of an isosceles triangle. If the roof angle is 48°, what acute angle does each side make with the ground?

 a. 42°
 b. 48°
 c. 54°
 d. 66°
 e. 132°

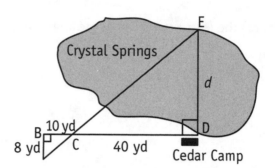

10. Using the drawing at right, determine the distance (*d*) across Crystal Springs at Cedar Camp.

 a. 24 yards
 b. 28 yards
 c. 32 yards
 d. 40 yards
 e. 48 yards

ΔABC and ΔCDE are similar triangles.

11. At 3 o'clock in the afternoon, Joyce measures the shadow of a yardstick. The yardstick, while being held vertically, casts a shadow of 4 feet. At the same time, a nearby flagpole casts a shadow of 32 feet. What is the height of the flagpole?

 a. 24 feet
 b. 27 feet
 c. 30 feet
 d. 32 feet
 e. 36 feet

12. Jesse hiked 12 miles east and 5 miles north of his car. What straight-line distance is Jesse from his car?

 a. 7 miles
 b. 10 miles
 c. 12 miles
 d. 13 miles
 e. 17 miles

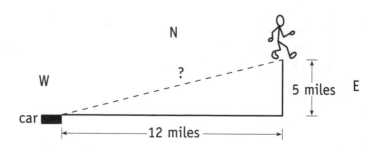

13. The flower display at the right is in the shape of a right triangle. What is the length of side AB?

 a. 3 feet
 b. 5 feet
 c. 6 feet
 d. 8 feet
 e. 10 feet

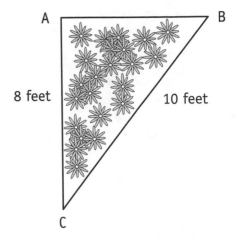

14. Leona placed 20 decorative bricks end to end in a straight line. Each brick measures 15 inches long. What is the total length of the line of bricks?

 a. 21 feet
 b. 22 feet
 c. 23 feet
 d. 24 feet
 e. 25 feet

15. Paul's bedroom is in the shape of a square. The distance around the outside edge of this room is 48 feet. How many square yards of carpet are needed to carpet Paul's bedroom? (**Hint:** First, find the length in yards of each side of the room.)

 a. 16 square yards
 b. 24 square yards
 c. 48 square yards
 d. 96 square yards
 e. 144 square yards

16. A corner table is in the shape of a right triangle. What is the area of the top of this table in square feet?

4 feet

2 feet

 a. 4 square feet
 b. 6 square feet
 c. 8 square feet
 d. 12 square feet
 e. 16 square feet

Questions 17 and 18 refer to the drawing at the right.

 17. Lakeview Community Center built the circular wading pool shown at right. To the nearest foot, what is the circumference of this pool? (Use 3.14 or $\frac{22}{7}$ for π.)

 a. 33 feet
 b. 48 feet
 c. 66 feet
 d. 99 feet
 e. 132 feet

Wading Pool

diameter = 21 feet

 18. To the nearest 10 square feet, what is the area of the wading pool?

 a. 230 square feet
 b. 260 square feet
 c. 290 square feet
 d. 320 square feet
 e. 350 square feet

19. How many square feet of wallpaper would be needed to cover the area of wall shown at the right? (No wallpaper is needed for the windows!)

 a. 81 square feet
 b. 138 square feet
 c. 150 square feet
 d. 162 square feet
 e. 186 square feet

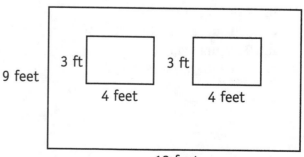

9 feet

3 ft 3 ft

4 feet 4 feet

18 feet

20. What is the storage volume of a freezer that has inside dimensions of 5 feet high, 2 feet wide, and $1\frac{3}{4}$ feet deep?

 a. $8\frac{3}{4}$ cubic feet

 b. $11\frac{3}{4}$ cubic feet

 c. $14\frac{1}{2}$ cubic feet

 d. $15\frac{3}{4}$ cubic feet

 e. $17\frac{1}{2}$ cubic feet

21. Use the volume formula for a cylinder, $V = \pi r^2 h$, to find the volume of the oil drum shown at the right. Express your answer to the nearest cubic foot.

 a. 6 cubic feet
 b. 16 cubic feet
 c. 20 cubic feet
 d. 28 cubic feet
 e. 37 cubic feet

radius = 1.25 feet

4 feet

22. How many cubic inches are in 1 cubic foot?

 a. 6 cubic inches
 b. 12 cubic inches
 c. 36 cubic inches
 d. 144 cubic inches
 e. 1,728 cubic inches

23. How many *cubic yards* of topsoil are needed for a garden that measures 45 feet long and 30 feet wide if the topsoil is to be 6 inches deep?

 a. 15 cubic yards
 b. 25 cubic yards
 c. 150 cubic yards
 d. 675 cubic yards
 e. 8,100 cubic yards

24. How many cubic inches of metal are contained in the hollow rectangular tube shown at the right?

a. 330 cubic inches
b. 380 cubic inches
c. 400 cubic inches
d. 465 cubic inches
e. 480 cubic inches

Geometry Posttest B Chart

Circle the number of any problem that you miss and be sure to review the appropriate practice pages. A passing score is 20 correct answers. If you passed the test, go on to Using Number Power. If you did not pass the test, take the time to make a more thorough review of the book.

PROBLEM NUMBERS	SKILL AREA	PRACTICE PAGES
1	measurement of angles	16–17
2	types of angles	18–19
3, 5	pairs of angles	26–27
4	pairs of angles: applying skills	28–29
6, 7	triangle: sum of angles	44–45
8, 9	triangle: applying skills	46–47
10	similar triangles	52–59
11	similar triangles: applying skills	60–61
12, 13	Pythagorean theorem: applying skills	68–69
14	U.S. customary length units	78–79
15	square: perimeter + area	94–97
16	triangle: area	104–105
17	circle: circumference	110–111
18	circle: area	112–113
19	two-step area problem	118–121
20	rectangular solid: volume	133–134
21	cylinder: volume	135
22	cube: volume	131–132
23	volume: applying skills	137–138
24	two-step volume problem	139–142

Using
Number
Power

The Work Triangle

Probably no room in your house is used more than your kitchen. In fact, the kitchen is such an important room that studies have been done to identify what the characteristics of an ideal kitchen might be.

Home economists have noted that the kitchen can be divided into three main work areas: the sink, the stove, and the refrigerator. The most practical and efficient kitchen is one in which these three work areas are placed so that they form the vertex points of a triangle. This triangle is often called the **work triangle.**

Ideally, the distances between work areas should be as shown below.

Note: Work areas should not be closer together than 4 feet.

The perimeter of the work triangle is a measure of kitchen efficiency. For a very efficient kitchen, this perimeter should be no more than 22 feet. Also, for a person working alone in the kitchen, the most efficient work triangle is an equilateral triangle with each side measuring 4 feet.

1. **a.** What is the perimeter of the work triangle of the kitchen pictured at the right?

 b. Would you say this is a kitchen correctly designed for work efficiency?

The work triangle (dotted line) joins the front centers of the refrigerator, range, and sink.

2. **a.** What is the perimeter of the work triangle of the kitchen at the right?

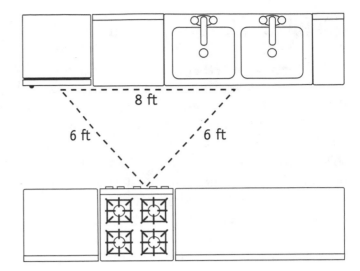

8 ft

6 ft 6 ft

b. Compare the perimeters of the work triangles to decide if this kitchen is as efficiently designed as the kitchen in problem 1.

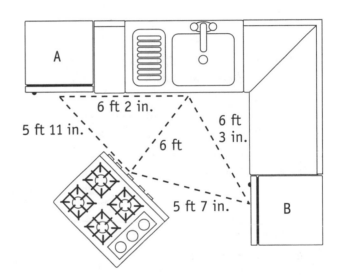

3. In the kitchen at the right, the refrigerator can be placed at either point A or point B. Which location gives the greatest work efficiency? Why?

A

6 ft 2 in.

5 ft 11 in.

6 ft 3 in.

6 ft

5 ft 7 in. B

4. Use the work triangle idea to design an ideal kitchen for one person.

 a. The ideal kitchen is drawn in the shape of what kind of triangle?

 Draw the kitchen here.

 b. What is the length of each side of the work triangle?

Applying the Pythagorean Theorem

The **Pythagorean theorem** is often used to find long distances. To use this theorem to solve the problems in this section, you need to know how to find or to approximate the square root of large numbers. Often, this is easily done by using a calculator. Another way is to write a table of perfect squares and approximate the square root you need to the nearest whole number.

For the problems below, use your calculator or the Partial Table of Perfect Squares at the right to approximate the square roots needed to solve each problem.

The symbol ≈ means "is approximately equal to."

Partial Table of Perfect Squares	
$24^2 = 576$	$71^2 = 5,041$
$25^2 = 625$	$72^2 = 5,184$
$26^2 = 676$	$73^2 = 5,329$
$44^2 = 1,936$	$100^2 = 10,000$
$45^2 = 2,025$	$101^2 = 10,201$
$46^2 = 2,116$	$102^2 = 10,404$

Find the approximate square root in each of the following problems. The first has been done as an example.

1. A power company employee is going to run a power line from the power pole to the back of Joel's house. Use the drawing at the right, and determine the approximate length of cable the employee will need.

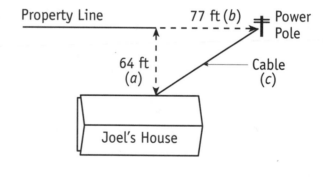

$c^2 = a^2 + b^2$
$c^2 = 64^2 + 77^2$
$c^2 = 4,096 + 5,929$
$c^2 = 10,025$
$c = \sqrt{10,025}$
Since 10,025 is closest to 10,000 on the table above, the approximate square root of 10,025 is 100.
c ≈ 100 ft

2. A forest ranger at Oak Ridge Lookout spotted a fire 24 miles west of his location. If the town of Olene is 38 miles due south, how far is Olene from the fire?

3. A disabled ship radios to shore for help. The Coast Guard determines that the ship is 16 miles east and 43 miles north of the station. What is the direct distance between the ship and the Coast Guard station?

4. Bill Adamly wants to hike directly from Mt. Scott to Mt. St. Clair. In mountain terrain, he can average about one mile for each hour he hikes. Use the map at the right to estimate how long it will take Bill to hike between the two mountains.

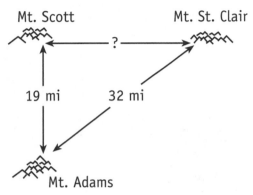

5. Jackson is 54 miles from Lazy R Resort. Ontario is 31 miles south of Jackson. A land developer proposes building a shortcut road to directly connect Ontario and Lazy R. Use his drawing at the right to estimate the length of this new road.

6. Radio station KNPG in Bingham has a broadcast range of 70 miles. Refer to the drawing at the right and determine if Ellensburg is within the broadcast range of KNPG.

7. Eva Lewis wants to put an underground sprinkler system in her backyard. A drawing of the system is shown at the right. Estimate how many feet of water pipe Eva will need. (**Hint:** The unmeasured length of pipe is equal to the hypotenuse of a right triangle.)

Tiling a Room

Tiling a room is a home-improvement activity that many people do themselves. No particular professional skills are needed, and tiles can be laid in a short time.

There are three main types of tile available. Although costs vary with location, here are average prices.

Type of Tile	Cost per Tile	Size of Tile	Use of Tile
Self-stick vinyl	$0.75 to $1.50	12″ by 12″square	Use over old tile or linoleum.
Vinyl composition*	$0.90 to $2.50	12″ by 12″ square	Use on new (particle board or concrete) floor where average wear is expected.
Pure vinyl*	$3.00 to $4.50	9″ by 9″ square	Use on floor where very tough wear is expected.

*Mastic (tile glue) is required. Mastic costs about $7.50 a quart (to cover 75 sq ft).

Follow these steps to compute the cost of tiling a room.

STEP 1 Find the area of the floor to be tiled.

STEP 2 Find the number of tiles needed by dividing the floor area by the area of each tile.

STEP 3 Multiply the number of tiles needed by the cost per tile.

STEP 4 Add the cost of mastic if it is needed.

1. Lila Andrews wants to tile the living room shown below. She chooses self-stick vinyl tile that costs $0.80 for each 12″ by 12″ (1 sq ft) tile.

 a. How many tiles does Lila need for the living room?

 b. What is the total cost of this tile?

 c. Lila also considers buying tile at a close-out sale for $0.50 per 12″ by 12″ tile. However, she would have to buy 350 tiles at this price. What would be the cost of these 350 close-out tiles?

 d. How many close-out tiles would Lila have left over after tiling the living room?

Living Room 16 ft

13 ft

2. Lila also decides to tile the family room shown at the right. She buys tiles at a close-out sale for $.50 per 12″ by 12″ tile. However, to buy the tile at this low price, she has to buy 250 tiles.

 a. What is the total cost of the tile?

 b. After tiling the family room, how many tiles will Lila have left over?
 (**Note:** There are no tiles under the shelves.)

3. George Lawson plans to put vinyl-composition tile on his shop floor. The tile he's chosen costs $1.25 per 9″ by 9″ tile.

 a. What is the area of the shop floor to be tiled? The cabinet area is not tiled.

 b. How many 9″ by 9″ tiles are needed?
 (**Hint:** Since 9″ = $\frac{3}{4}$ ft, a tile that is 9″ by 9″ is $\frac{3}{4} \times \frac{3}{4} = \frac{9}{16}$ sq ft.)

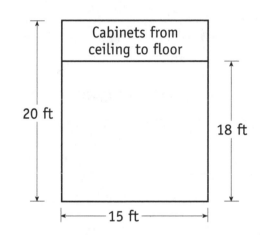

 c. What is the total cost of tiles?

 d. Mastic must be applied to the floor before vinyl-composition tile is laid. If each quart can of mastic covers about 75 sq ft, how many cans of mastic are required?

 e. What is the total cost of the mastic?

 f. What is the total cost of the tiles plus the mastic?

Building a Patio

Although it requires some work, many people build their own patios. In doing so, they save about half the cost of having a patio built for them.

A simply built patio can be made from a 4-inch thick slab of concrete. You first make forms out of two-by-fours to contain the wet concrete, and then you have ready-to-pour concrete delivered and poured directly into the forms. As the concrete is poured, you smooth it out to fill the forms. These forms are removed when the concrete has hardened.

The cost of delivered ready-to-pour concrete varies with location, but for small quantities an average price is about $90.50 per cubic yard. To find the cost of building your own concrete patio, multiply the number of cubic yards of concrete needed by the price of concrete per cubic yard.

1. Determine the cost of concrete required for the patio at the right by answering the following questions.

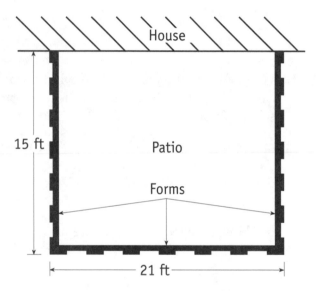

 a. What are the dimensions of the patio in yards?
 (**Hint:** To change feet to yards, divide by 3. To change inches to yards, divide by 36.)

 Length = 21 ft = _____ yd

 Width = 15 ft = _____ yd

 Thickness = 4 in. = _____ yd

 b. What is the volume, in cubic yards, of concrete needed for the patio? (Round up to the nearest cubic yard.)

 c. If concrete is $90.50 per cubic yard, what is the cost of the concrete?

 d. What would be the additional cost of covering the finished patio with a ceramic tile that costs $2.50 per square foot?

2. George is saving $40.00 a month to pay for a patio he wants to build himself. A drawing of the patio is at the right.

 a. How many cubic yards of concrete will be needed for this 4″ thick patio? (Round up to the nearest cubic yard.)

 b. At $85.50 per cubic yard, how much will the concrete cost?

 c. How many months must George save to have enough money to pay for the concrete?

3. Janet is building herself a patio as shown at the right. The patio is in the shape of a half circle and is 4 inches thick. (In these problems, use 3.14 for π.)

 a. To the nearest cubic yard, how much concrete will Janet need for her patio?

 b. At $88.00 per cubic yard, what will she pay for this concrete?

 c. Janet also wants to put a fancy brick edge along the circular perimeter of the patio. If each brick is 1 foot long, about how many bricks will she need?

 d. At $1.49 per brick, how much will she pay for the brick perimeter?

Carpeting a Room

Carpet is sold by the square yard. The price usually includes the cost of installation and the cost of a carpet pad, a foam rubber pad that is put on the floor under the carpet.

Cutting a carpet to fit a room and installing the carpet require professional skills and tools. Because of this, the cutting and installing of carpet is usually done by personnel from the store where you purchase your carpet.

When you buy carpet, you usually have to buy a piece that is larger than the room you are carpeting. This is because carpet comes in a 12-foot wide roll. You can have it cut to any length you want, but you usually must buy a 12-foot wide piece.

To find the cost of carpeting a room, multiply the number of square yards of carpet to be purchased times the price per square yard.

1. Figure out the cost of carpeting the bedroom at the right by answering the following questions.

 a. What are the dimensions of the room in yards? (**Hint:** To change feet to yards, divide by 3.)

 Length = 13 ft = _____ yd

 Width = 12 ft = _____ yd

 b. What is the area, in square yards, of carpet needed for this room?

 c. At $26.00 per square yard for carpet, pad, and installation, what is the cost of carpeting this bedroom?

Because the bedroom is 12 ft wide, a piece of carpet can be purchased that is the same size as the room.

2. How much will Mary pay to have her daughter's room carpeted if she pays $28.50 per square yard for carpet, pad, and installation?

3. To carpet the dining room at the right, Charles must buy a piece of carpet that is larger than the room itself. Refer to the drawing and answer these questions.

 a. How many square yards of carpet must Charles buy to carpet the room?

Size of carpet purchased.

 b. At $27.00 per square yard for carpet, pad, and installation, what is the cost of carpeting this room?

 c. How many square feet of carpet will Charles have left over?

Note: Because carpet can only be purchased in a 12-foot width, a piece 11 ft by 12 ft must be purchased to carpet a room that measures 11 ft by 11 ft. The carpet is cut to fit the room, and the leftover piece belongs to the buyer.

4. a. What is the cost of carpeting the dining room at the right if the cost of carpet, pad, and installation is $24.00 per square yard?

 b. How many square feet of carpet will be left over?

Yard Work

Lucy Boyd plans to make some improvements in her backyard. She wants to put a circular rose garden in the backyard and a vegetable garden behind the yard. She also plans to put a picket fence around the rose garden and a cedar fence around the vegetable garden. To enrich the grass, she is going to use lawn fertilizer.

To help estimate costs, she drew the following picture of her backyard.

 Using the drawing above, answer the following questions to help estimate the costs of Lucy's yard improvement ideas.

LAWN COSTS

1. Lucy plans to fertilize the lawn. Lawn fertilizer costs $5.80 a bag, and each bag holds enough fertilizer to cover about 1,500 square feet of lawn.

 a. Not counting the rose and vegetable gardens, what is the area of backyard lawn that Lucy needs to fertilize?

 b. If Lucy also has 4,600 square feet in her front and side lawns, how many bags of fertilizer will she need for the entire lawn?

 c. What will be the total cost of fertilizer?

ROSE GARDEN COSTS

2. Lucy wants to plant individual rose bushes in her garden. The salesperson tells her that for the variety of rose she wants, she should allow about 14 square feet of garden space for each plant.

 a. About how many rose bushes can Lucy plant in her circular rose garden?

 b. If each rose bush costs $6.95, what will be the total cost of the rose bushes?

 c. The decorative picket fence she's chosen comes in 1-foot sections. At $1.39 per 1-foot section, what will be the total cost of picket fence material?

VEGETABLE GARDEN COSTS

3. Because her soil quality is poor, Lucy decides to add a 4-inch layer of organic topsoil to her vegetable garden.

 a. What is the area of the vegetable garden space?

 b. At $22 per cubic yard, what will be the total cost of topsoil? (**Hint:** When spread to a depth of 4 inches, 1 *cubic yard* of topsoil will cover about 80 *square feet* of garden space.)

 c. Lucy also wants to fence the vegetable garden. The fence will enclose the garden except for a 3-foot gate. At $2.49 per foot, approximately how much will the cedar fence material cost Lucy?

Changing Volume to Capacity

In our study of geometry, we have referred to the space taken up by an object as its volume. When you measure the amount of liquid that a container will hold, you give that volume the special name, **capacity.** In the U.S. customary system, familiar units of capacity are the cup, pint, quart, and gallon. In the metric system, they are the liter and milliliter. Capacity units are directly related to the common volume units you have been using.

In many occupations, it is necessary to measure the volume of space inside an object and then to convert this measurement to capacity units. On these two pages, you'll look at the most common example: converting cubic feet to gallons.

For the example and the problems, you need to know that 1 cubic foot = $7\frac{1}{2}$ gallons.

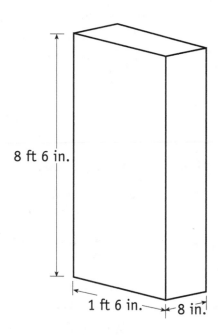

8 ft 6 in.

1 ft 6 in.　8 in.

EXAMPLE How many gallons of water will the container shown at the right hold?

STEP 1 Change each distance to feet only.

length: 1′ 6″ = $1\frac{1}{2}$ ft

width: 8″ = $\frac{2}{3}$ ft

height: 8′ 6″ = $8\frac{1}{2}$ ft

STEP 2 Find the volume in cubic feet by multiplying length times width times height. To multiply, express each distance as an improper fraction and cancel where possible.

$V = lwh$

$V = 1\frac{1}{2} \times \frac{2}{3} \times 8\frac{1}{2}$

$V = \frac{\cancel{3}^{1}}{\cancel{2}_{1}} \times \frac{\cancel{2}^{1}}{\cancel{3}_{1}} \times \frac{17}{2} = \frac{17}{2}$ or $8\frac{1}{2}$ cubic feet

STEP 3 To find the number of gallons in the container, multiply the number of cubic feet ($8\frac{1}{2}$) by the number of gallons in each cubic foot ($7\frac{1}{2}$).

Number of gallons = $8\frac{1}{2} \times 7\frac{1}{2}$

$= \frac{17}{2} \times \frac{15}{2} = \frac{255}{4} = \mathbf{63\frac{3}{4}}$

ANSWER: The container will hold $63\frac{3}{4}$ gallons of water.

1. Arny works in a sheet metal shop and makes custom built containers. He has an order for a custom built gasoline tank as shown at right. How many gallons of gas will this tank hold?

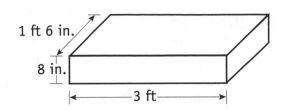

2. Dave's Auto Shop specializes in engine tune-up. Dave averages 7 oil changes a day, and he stores the used oil in the barrel shown at the right. If an oil change averages 1 gallon of used oil, how many days does it take Dave to fill the storage barrel?

3. A waterbed mattress is shown at the right. To determine the weight of this mattress when it is full of water, answer these questions.

 a. How many cubic feet of water does the mattress hold when full?

 b. How many gallons of water is this?

 c. If water weighs about $8\frac{1}{3}$ pounds per gallon, what is the weight of water in the full waterbed?

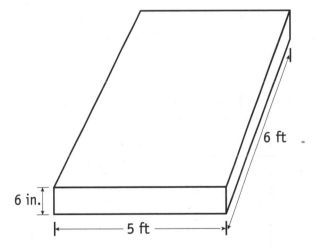

4. Dorothy wants to design a water tank that will hold at least 100 gallons of water. Does the tank she designed, shown at the right, meet her requirement?

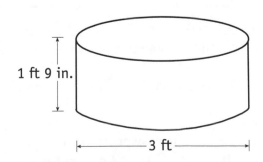

Design Work

As part of her job, Mai Lin helps design cardboard boxes.

DESIGNING FOR MAXIMUM VOLUME

Mai Lin's first task is to design a small open box for Tot Toys. Mai Lin is to start with a flat rectangular piece of cardboard that is 18 inches long and 11 inches wide. The box is made by cutting a small square from each corner of the cardboard and folding the sides up along the dotted lines as shown. The box must have as much volume as possible, but it will not have a top.

For example, suppose a 1-inch by 1-inch square is cut out of each corner. The bottom will now be 16 inches long (18 − 2), 9 inches wide (11 − 2), and 1 inch high. After folding on the dotted lines, the volume will be 16 × 9 × 1 = 144 cubic inches.

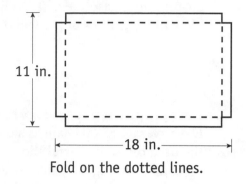

Fold on the dotted lines.

 1. Complete the table to determine the volume of the box for each square size indicated.

Side of Square	Length of Box	Width of Box	Height of Box	Volume of Box
1 in.	16 in.	9 in.	1 in.	140 cu in.
1.5 in.	15 in.	8 in.	1.5 in.	
2 in.				
2.5 in.				
3 in.				
3.5 in.				
4 in.				
4.5 in.				
5 in.				

2. **a.** To the nearest half inch, what is the size of the square that gives the box the maximum volume?

 b. What is the maximum volume?

3. Does your answer tell the true maximum volume possible or is it an estimate? Give a reason for your answer.

DESIGNING FOR MINIMUM MATERIAL

Mai Lin's second task is to decide which of two boxes to use for packaging sand that will be sold by Bay Aquarium Products. The two boxes she must choose between are shown below.

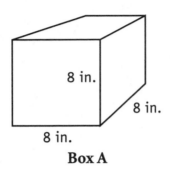

8 in.

8 in.

8 in.

Box A

4 in. 8 in.

16 in.

Box B

4. What volume of sand will each box hold?

 Box A _____

 Box B _____

5. Which box do you think takes more cardboard to make? (Consider just the surface area of each box.)

 Box A _____ _____ takes more cardboard.

 Box B _____

6. Which box do you think Mai Lin should choose? Give a reason for your answer.

Tessellations

A tessellation is a pattern that is formed by repeating a single shape in a side-by-side manner and filling all the available space. A honeycomb is an excellent example. There are no gaps in a tessellation. The shapes fit snuggly together.

Tessellations are used in many areas of design. You see tessellations in the design of floor tiles, carpets, curtains, quilts, clothing, brick sidewalks, and buildings.

Notice that not all shapes form tessellations. Circles, for example, leave gaps when placed side by side.

Honeycomb Tessellation

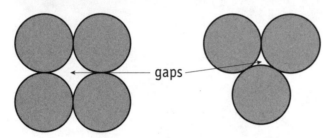 gaps

1. Describe the geometric shape (hexagon) that is repeated in a honeycomb.

2. Draw a partial tessellation using each of the following shapes.

Triangle Rectangle Trapezoid

3. Can you form a tessellation using a pentagon? A pentagon is a 5-sided regular polygon: 5 equal sides and 5 equal angles. (**Hint:** Trace several like the one at the right. Cut them out and place them side by side.)

Drawing 2-Dimensional Views of 3-Dimensional Objects

An applied geometry skill used in design is the drawing of different views of an object. An important example is drawing a 2-dimensional view of a 3-dimensional shape.

EXAMPLE Three sugar cubes are placed side by side to make the figure shown. The front, right-side, and left-side views of the figure are also shown. To determine these views, visualize what the figure looks like from each view direction.

Figure

Front View

Right-side View

Left-side View

Top View

1. The figure at the right is made out of six cubes. On the grids below, draw each view indicated.

Front View

Top View

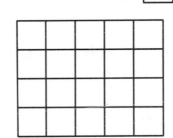

Left-side View

2. A figure is made out of eight cubes. The front and right-side views of the figure are shown at the right.

Front View

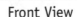

Right-side View

a. Draw the left-side view.

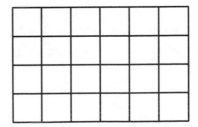

b. Draw the top view.

Precision in Measurement

In the workshop, measurement is not exact. In fact, finished dimensions are always given as rounded numbers. For example, it is not possible to cut wood or metal to an exact length or width. Suppose you are asked to cut a piece of wood to be exactly 3 feet wide. This is not possible. You may be able to cut the wood, though, to be 3 feet wide *to the nearest* $\frac{1}{32}$ *inch*. In measurement, the word *precision* describes how close a finished dimension is to the desired exact dimension.

Each dimension of the rectangular piece of brass shown at the right is *rounded to the nearest inch*.

3 in.

7 in.

1. **a.** What is the *least* length and width the brass may actually have? length _____

 width _____

 b. What is the *greatest* length and width the brass may actually have? Give each answer to the nearest $\frac{1}{8}$ inch. length _____

 width _____

2. **a.** Using your answers from problem 1a, find the *least area* the rectangle may actually have.

 b. Using your answers from problem 1b, find the *greatest area* the rectangle may actually have.

3. Using your answers from problem 2, determine the *range* in the possible area values. (Range = greatest possible value − least possible value)

Dimensions of a Sphere

A **sphere** is a 3-dimensional shape you're very familiar with. A basketball has the shape of a sphere. So do a tennis ball and a baseball.

A sphere, like other 3-dimensional shapes, has both surface area (S) and volume (V), given by these formulas:

Surface area: $S = 4\pi r^2$ Volume: $V = \frac{4}{3}\pi r^3$

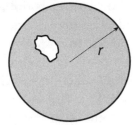

Sphere

EXAMPLE What is the surface area and volume of a basketball that has a radius of 5 inches?

Surface area	**Volume**

$S = 4\pi r^2$

$\approx 4(3.14)5^2$

$= 4(3.14)(25)$

$= 314$ sq in.

$V = \frac{4}{3}\pi r^3$

$\approx \frac{4}{3}(3.14)5^3$

$= \frac{4}{3}(3.14)(125)$

$= 523$ cu in.

 1. What is the surface area and volume of a small ball that has a Surface area: _____
radius of 1 inch?

Volume: _____

 2. The radius of the earth is about 4,000 miles.

a. To the nearest thousand miles, what is the approximate circumference (C) of the earth? (**Hint:** $C = \pi d$)

b. To the nearest million square miles, what is the approximate surface area of the earth?

c. If the surface of the earth is about 30% land, approximately how many million square miles of land are on the surface of the earth?

d. If the surface of the earth is about 70% water, approximately how many million square miles of water are on the surface of the earth?

Using Geometry in Photography

Photography is the process of recording an image on film placed inside a camera. Prints (commonly called pictures) are then made from the image that appears on the processed film—called a negative.

An image is produced on film when light from the subject being photographed passes through a lens and exposes the film. Using geometry and a simple camera, we can understand the relation between the height of the subject and the height of the image.

In the drawing above, the subject S is being photographed. The **image** I appears on the film. The **image distance** CD is commonly called the **focal length.**

Triangles ABC and CDE are similar triangles and are formed by rays of light moving from the subject, through the lens, and to the film.

EXAMPLE In the camera above, the image distance (focal length CD) is 2 inches. What is the height of the image on the film if the subject is a 50 foot high tree that is 200 feet from the camera?

STEP 1 Convert all distances to inches.

AB = 50 feet = 50 × 12 = 600 inches
CD = 2 inches
BC = 200 feet = 200 × 12 = 2,400 inches

STEP 2 Write a proportion.

$$\frac{\text{image height}}{\text{subject height}} = \frac{\text{image distance}}{\text{subject distance}}, \text{ or } \frac{DE}{AB} = \frac{CD}{BC}$$

where DE = I, the image height you are trying to find.

Thus, $\dfrac{I}{600} = \dfrac{2}{2,400}$

STEP 3 Solve for I.

a. Cross multiply.

$2,400 \times I = 2 \times 600$
$2,400\, I = 1,200$

b. To find I, divide 1,200 by 2,400.

$I = \dfrac{1,200}{2,400} = \dfrac{1}{2}$

ANSWER: The size of the image on the film is $\frac{1}{2}$ inch.

Use $\dfrac{\text{image height}}{\text{subject height}} = \dfrac{\text{image distance}}{\text{subject distance}}$ to solve each problem.

1. Using a lens with a 1-inch focal length (image distance), what is the height of the image when the subject is a 200-foot-high building that is 500 feet away?

2. Sarah uses a 35 mm camera with a lens that has a focal length of 2 inches. For this lens, the maximum image height is 1 inch.

 a. If Sarah photographs her 5 foot tall sister, how far should Sarah stand from her so that the image will be the maximum 1 inch?

 b. What will be the height of the image if Sarah stands 25 feet from her sister?

3. Using a zoom lens, a photographer can change the focal length over a wide range of values. A zoom lens allows you to change the image size without changing your distance from the subject.

 a. If you wish to photograph a 125-foot-high tree from the other side of a 875-foot-wide canyon, what focal length is needed to produce a 1-inch-high image?

 b. Taking a picture at this same distance, what would be the size of the image if you changed the focal length to 5 inches?

Geometric Illusions

In your study of geometry, you have learned about the measurement of angles, length, area, and volume. Another interesting topic in geometry is **geometric illusions.**

A geometric illusion is produced when one or more geometric figures appears distorted or different than it really is due to the presence of other figures. You'll never be tested on your knowledge of geometric illusions, but they are fun to look at. You'll see several well-known illusions on these two pages.

LINE-LENGTH ILLUSIONS

In each drawing below, circle the line in each pair that looks longer. Use a ruler to check each answer.

1.

AB or CD

2.

a or b

3.

i or j

4.

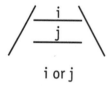

AB or BC

RIGHT-ANGLE ILLUSIONS

There are two right angles in the drawing that follows. See if you can spot them without measuring. Check your answers with a protractor or with the corner of a piece of paper.

5.

PARALLEL-LINE ILLUSIONS

In each drawing below, do the two horizontal center lines bow in, bow out, or are they parallel? Use a ruler to check the actual distance between the lines in each problem.

6.

7.

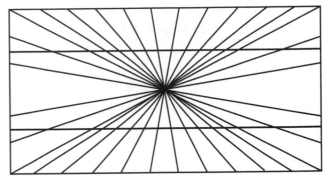

AREA ILLUSIONS

In each drawing below, choose the circle that has the largest area.

8.

c or d

9.

dotted R or dotted S

VOLUME ILLUSIONS

Which of the two rectangular solids drawn below has the greatest volume?

10.

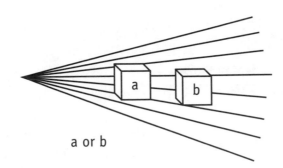

a or b

Using a Calculator

A calculator is a valuable math tool. You'll use it mainly to add, subtract, multiply, and divide quickly and accurately. The calculator pictured at the right is similar to one you've seen or may be using.

To enter a number, press one digit key at a time. On the display at the right, the number 4,863 is entered. Notice the following features:

- Display
- Memory Keys
- Function Keys
- Digit Keys
- Clear Keys

- A decimal point is displayed to the right of the ones digit.
- A comma is not used to separate groups of digits.
- The calculator does not have a comma (,) key or a dollar sign ($) key.
- Pressing the clear key (**C**) erases the display. You should press **C** each time you begin a new problem or when you've made a mistake.

__EXAMPLE 1__ Divide 408 by 12.

> **STEP 1** Press **C** to clear the display.
>
> **STEP 2** Press the digit keys, divide key (÷), and equals key (=).

Press Keys 　　　　　　　　　　**Display Shows**

　C　4　0　8　÷　1　2　=　　　　　34.

ANSWER: 34

__EXAMPLE 2__ Multiply $12.75 by 6.

> **STEP 1** Press **C** to clear the display.
>
> **STEP 2** Press the digit keys and decimal point key, times key (×), and equals key (=).

Press Keys 　　　　　　　　　　**Display Shows**

　C　1　2　.　7　5　×　6　=　　　　76.5

ANSWER: $76.50

↑ Press the decimal point key to separate dollars from cents.

↑ Calculators do not show zeros at the right end of a decimal fraction.

Using Mental Math

Mental math is the use of thinking skills to help you understand and solve math problems. Estimating in your head is one mental math skill you're already familiar with. As you know, estimating is useful when adding, subtracting, multiplying, and dividing. Another mental math skill especially valuable in *Geometry* is being familiar with geometric relationships—among angles, figures, and type of geometric measurement (perimeter, area, and volume). Here are some key ideas to help build your mental math skills in these areas.

ANGLES

Angles can be added and subtracted like numbers. Adding two angles results in a larger angle. Subtracting a smaller angle from a larger angle results in a second small angle. You can estimate angles the same way you estimate numbers. For example, you can estimate the sum of a 78° angle plus a 53° angle as follows: $78° + 53° ≈ 80° + 50° = 130°$.

In this book you learned a very important rule about triangles: the sum of the three angles in a triangle is always 180°. This fact tells you several more things about triangles.

- A triangle can have only one 90° angle.
- A triangle can have only one obtuse angle (an angle larger than 90°).
- As one angle in a triangle increases, the sum of the other two angles decreases.

See if you can tell why each of these statements is true!

FIGURES

Here is a fun game that will help develop your intuition about geometric figures. Get six identical toothpicks. By placing different numbers of toothpicks end-to-end, determine what types of figures (shapes) you can make. Describe each figure as completely as you can. Here are three example questions you may find interesting.

- Can you make a triangle using four toothpicks? Tell why or why not.
- How many different types of figures can you make using four toothpicks? Name them.
- Can you make a circle using toothpicks? Tell why or why not.

PERIMETER, AREA, AND VOLUME

To strengthen your skills with perimeter, area, and volume, remember these facts:

- Perimeter is a 1-dimensional measurement. (Only one length is involved.)
- Area is a 2-dimensional measurement. (Multiplying two lengths is always involved.)
- Volume is a 3-dimensional measurement. (Multiplying three lengths is always involved.)

There is an important number in math whose symbol is $π$, pronounced "pie." The value of $π$ is about 3.14 or $\frac{22}{7}$. Remember this: The value $π$ is used only with circles. Any perimeter, area, or volume that involves $π$ alerts you that a circle is involved. And, any question that asks about a figure in which a circle is involved will require you to use $π$. Circles and cylinders both involve $π$. (See Formulas, pages 187–188.)

Using Estimation

To estimate is to compute an approximate answer. One way to estimate that works well is to use rounded numbers that are easy to work with. Estimating is an important skill for several reasons. You can estimate to

- discover about what an exact answer should be
- help pick a correct answer from among multiple choices—such as on multiple choice questions found on many tests
- quickly check an answer obtained when using a calculator—making sure that you worked the problem correctly *and* that you did not make a keying error

In *Geometry* there are questions for which you are asked to estimate an answer. On these questions, you do not need to calculate an exact answer. Use your estimation skills to discover an approximate answer. Here are some general guidelines for estimating answers to problems you'll find in this book.

WHOLE NUMBERS

To estimate with whole numbers, replace each number by a rounded number that contains one or more ending zeros.

EXAMPLE 1 Multiply. 198×42
Estimate. $200 \times 40 = \textbf{8,000}$

> Round 198 to the nearest hundred; round 42 to the nearest 10.

MIXED NUMBERS

Round mixed numbers by replacing each mixed number with the nearest whole number.

EXAMPLE 2 Add. $7\frac{7}{8} + 3\frac{1}{4} + 4\frac{1}{2}$
Estimate. $8 + 3 + 5 = \textbf{16}$

> Round $7\frac{7}{8}$ to 8; $3\frac{1}{4}$ to 3; and $4\frac{1}{2}$ to 5.

MIXED DECIMALS

Round mixed decimals by replacing each mixed decimal with the nearest whole number.

EXAMPLE 3 Multiply. 12.85×4.2
Estimate. $13 \times 4 = \textbf{52}$

> Round 12.85 to 13; round 4.2 to 4.

FORMULAS

To estimate with formulas, round whole numbers, mixed numbers, and mixed decimals as discussed above. Round π to 3 for a quick calculation.

EXAMPLE 4 Multiply. $\pi \times 6\frac{7}{8} \times 6\frac{7}{8}$
Estimate. $3 \times 7 \times 7 = \textbf{147}$

> Round π to 3; round each $6\frac{7}{8}$ to 7.

Formulas

PERIMETER

Figure	Name	Formula	Meaning
	Rectangle	$P = 2l + 2w$	l = length w = width
	Square	$P = 4s$	s = side
	Triangle	$P = s_1 + s_2 + s_3$	s_1 = side 1 s_2 = side 2 s_3 = side 3
	Polygon (n sides)	$P = s_1 + s_2 + \ldots s_n$	s_1 = side 1, etc.
	Circle (circumference)	$C = \pi d$	$\pi \approx 3.14$ or $\frac{22}{7}$ d = diameter

AREA

Figure	Name	Formula	Meaning
	Rectangle	$A = lw$	l = length w = width
	Square	$A = s^2$	s = side
	Parallelogram	$A = bh$	b = base h = height
	Trapezoid	$A = \frac{1}{2}(b_1 + b_2)h$	b_1 = base 1 b_2 = base 2
	Triangle	$A = \frac{1}{2}bh$	b = base h = height
	Circle	$A = \pi r^2$	$\pi \approx 3.14$ or $\frac{22}{7}$ r = radius
	Sphere (surface area)	$A = 4\pi r^2$	$\pi \approx 3.14$ or $\frac{22}{7}$ r = radius

VOLUME

Figure	Name	Formula	Meaning
	Rectangular solid	$V = lwh$	l = length w = width h = height
	Cube	$V = s^3$	s = side
	Cylinder	$V = \pi r^2 h$	$\pi \approx 3.14$ or $\frac{22}{7}$ r = radius h = height
	Cone	$V = \frac{1}{3}\pi r^2 h$	$\pi \approx 3.14$ or $\frac{22}{7}$ r = radius h = height
	Sphere	$V = \frac{4}{3}\pi r^3$	$\pi \approx 3.14$ or $\frac{22}{7}$ r = radius

ANGLES

Name	Figure	Formula	Meaning
Complementary angles		$\angle a + \angle b = 90°$	a and b are any two angles that add to 90°
Supplementary angles		$\angle a + \angle b = 180°$	a and b are any two angles that add to 180°
Sum of angles in a triangle		$\angle a + \angle b + \angle c = 180°$	a, b, and c are the three angles of any triangle
Pythagorean theorem		$c^2 = a^2 + b^2$	c = hypotenuse a and b are the two shorter sides of a right triangle

Glossary

A

acute angle An angle measuring more than 0° but less than 90°.

acute angle

angle Figure formed when two rays are joined, forming a point called a vertex of the angle.

Angles

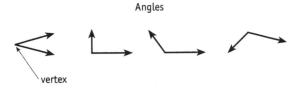

vertex

arc A small curved arrow ⌣, often used to indicate an angle.

area (*A*) An amount of surface

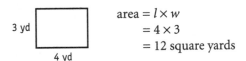

3 yd [] 4 yd

area = *l* × *w*
= 4 × 3
= 12 square yards

area unit Square units that are used to measure area. Common area units are square inches, square feet, square yards, and square meters.

area unit

C

capacity The volume of liquid (or substance such as sugar) that a container can hold. Common capacity units are fluid ounce, quart, gallon, milliliter, and liter.

centimeter A meter unit of length equal to 0.01 meter. 1 inch = 2.54 centimeters.

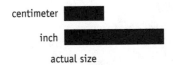

centimeter

inch

actual size

circle A plane (flat) figure, each point of which is an equal distance from the center.

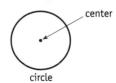

center

circle

circumference The distance around a circle. Circumference = π × diameter (π ≈ 3.14)

circumference

complementary angles Angles whose sum is 90°.

B

A

∠A and ∠B are complementary angles.

cone A 3-dimensional figure with a circular base and a vertex at a distance (height) across from the center of the base.

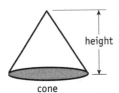

height

cone

congruent Having exactly the same size and shape.

congruent figures

corresponding angles Pairs of equal angles in similar triangles.

Similar Triangles

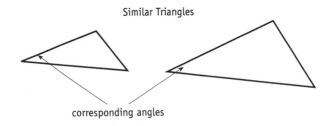

corresponding angles

corresponding sides In similar triangles, sides that are across from corresponding angles.

Similar Triangles

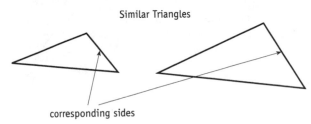

corresponding sides

cube A 3-dimensional shape that contains six square faces. At each vertex, all sides meet at right angles.

cube

cylinder A 3-dimensional shape that has both a circular base and a circular top.

cylinder

D

degrees A measure of the size of an angle. A circle contains 360 degrees (360°). One-fourth of a circle contains 90°.

 1°

 90°

$\frac{1}{4}$ circle

diagonal A line segment running between two non-consecutive vertices of a polygon. A diagonal divides a square or rectangle into two equal triangles.

diagonal

diameter A line segment from one side of a circle to the other that passes through the center. The length of the diameter is the distance across the circle.

diameter

distance formula The formula $d = vt$ that relates distance, velocity, and time. In words, *distance* equals *velocity* times *time*. By rearranging the variables, you can write the velocity formula $v = \frac{d}{t}$ or the time formula $t = \frac{v}{d}$.

E

English system Also called U.S. customary system. The English system uses inches, feet, yards, gallons, and other familiar units that are not part of the metric system.

equilateral triangle A triangle with three equal sides and three equal angles, each measuring 60°. An equilateral triangle is a regular polygon.

equilateral triangle

F

fluid ounce The smallest common unit of capacity in the U.S. customary system. One cup contains 8 fluid ounces; 1 quart contains 32 fluid ounces.

foot A unit of length equal to 12 inches. There are 3 feet in 1 yard.

G

gallon A unit of capacity equal to 4 quarts. One gallon contains 128 fluid ounces.
1 cubic foot $\approx 7\frac{1}{2}$ gallons

H

hypotenuse In a right triangle, the side opposite the right angle.

hypotenuse

I

isosceles triangle A triangle in which two sides have the same length. The two angles opposite the equal sides have the same measure. The two equal angles are called *base angles*.

base angles

isosceles triangle

L

line A path of points that extend in two opposite directions. A line has no endpoints.

line

line segment Part of a line, having two endpoints.

line segment

K

kilometer A metric unit of length equal to 1,000 meters. 1 meter \approx 0.6 miles.

kilometer

mile

drawn to scale

M

meter A metric unit of length equal to 100 cm.
1 meter ≈ 39.6 inches, a little longer than 1 yard.

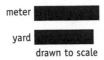

meter

yard

drawn to scale

metric system The measurement system used in most industrialized countries in the world. The United States is slowly adapting to metric measurement.

mile A length unit equal to 5,280 feet. One mile contains 1,760 yards.

O

obtuse angle An angle that measures more than 90° but less than 180°.

obtuse angle

P

parallel lines Equidistant lines that run side by side and do not cross.

parallel lines

parallelogram A four-sided polygon with two pairs of parallel sides. Opposite sides are equal and opposite angles have equal measure.

parallelogram

perfect square Whole numbers whose square roots are also whole numbers. For example, 49 is a perfect square: $\sqrt{49} = 7$.

perimeter The distance around a flat (plane) figure. The symbol for perimeter is P.

perpendicular lines Lines that meet (or cross) at a right angle (90°).

perpendicular lines

pi (π) The ratio of the circumference of a circle to its diameter. Pi is approximately equal to 3.14 or $\frac{22}{7}$.

plane figure A 2-dimensional (flat) figure. Examples are circles, squares, rectangles, and triangles.

plane figures

polygon A plane (flat) figure formed by three or more line segments that meet only at their endpoints. A *regular polygon* is a polygon in which all sides and all angles are equal.

polygons

pentagon

regular polygon

protractor A tool used for measuring and drawing angles.

Pythagorean theorem A theorem that states that in a right triangle the square of the hypotenuse is equal to the sum of the squares of the two remaining sides.

Pythagorean theorem

$c^2 = a^2 + b^2$

$5^2 = 4^2 + 3^2$

$25 = 16 + 9$

R

radius A line segment from the center of a circle to any point on the circumference of the circle.

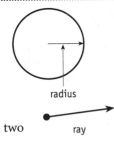

radius

ray Half of a line. A ray has one endpoint. An angle is formed when two rays are joined at their endpoints.

ray

rectangle A four-sided polygon with two pairs of parallel sides and four right angles. The opposite sides of a rectangle have equal length.

rectangle

rectangular solid (prism) A 3-dimensional figure in which each face is either a rectangle or a square. Opposite faces are congruent.

rectangular solid

reflex angle An angle that measures more than 180° but less than 360°.

reflex angle

right angle An angle that measures exactly 90°. A right angle is often called a "corner angle."

right angle

S

scalene triangle A triangle that has no equal sides and no equal angles.

scalene triangle

similar triangles Triangles that have three equal angles and the same shape. They differ only in the lengths of their sides.

similar triangles

sphere A 3-dimensional shape, each point of which is an equal distance from the center. A basketball is a sphere.

sphere

square A regular polygon with four equal sides, two pairs of parallel sides, and four right angles.

square

square of a number A second number formed by multiplying a first number by itself.
$8 \times 8 = 64 \leftarrow 64$ is the square of 8.

square root One of two equal factors of a number. The symbol for square root is $\sqrt{}$.
$\sqrt{9} = 3 \leftarrow 3$ is the square root of 9.

straight angle An angle that measures exactly 180°, having the shape of a straight line.

straight angle

supplementary angles Angles whose sum is 180°.

∠C and ∠D are supplementary angles.

T

tessellation A pattern made by repeating a figure in a side-by-side manner, leaving no gaps.

three-dimensional figures Figures that take up space and have volume. Examples are cubes, rectangular solids, and cylinders.

transversal A line that cuts across parallel lines, intersecting each of them.

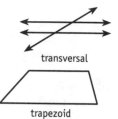
transversal

trapezoid A four-sided polygon with one pair of parallel sides.

trapezoid

triangle A three-sided polygon—a plane (flat) figure with three sides and three angles.

triangle

U

U.S. customary system The measurement system that uses inches, feet, yards, gallons, and other familiar units not part of the metric system.

V

vertex The point where the two sides of an angle meet. Vertex

vertex

vertical angles Angles formed when two lines cross. Vertical angles lie across from each other and are equal.

∠E and ∠F are vertical angles.

volume The amount of space an object takes up. Volume is usually measured in cubic units.

volume unit Cubic units that are used to measure volume. Common volume units are cubic inches, cubic feet, cubic yards, cubic centimeters, and cubic meters.

volume unit

Y

yard A unit of length equal to 36 inches or 3 feet.

yard

foot

inch

drawn to scale

Index

A

Angle(s) 14, 18, 26
 acute 18
 alternate exterior 30
 alternate interior 30
 complementary 26
 corresponding 30, 52
 labeling 20
 measuring 21
 obtuse 18
 reflex 19
 right 18, 64
 straight 18
 supplementary 26
 vertical 27
Area 92, 114
 circle 112
 parallelogram 108
 rectangle 100
 square 96
 triangle 104
 trapezoid 108
 two-step problem 118, 120
 units 92

C

Calculator 184
Capacity 172
Circle 110, 112
Circumference 110
Complementary angles 26
Cone 127, 136
Corresponding angles 30, 52
Corresponding sides 52
Cube 127, 131
Customary units 78
Cylinder 127, 135

D

Degrees 16
Diameter 110

E

Estimation 186
Equivalent fractions 54

F

Formulas 114, 137, 187

G

Geometric illusions 182

H

Hypotenuse 64

L

Length units 78, 80
Line 14

M

Measurements 78
Mental math 185
Metric units 80

P

Parallel lines 30, 32
Parallelogram 90, 106, 108
Perfect square 63
Perimeter 88, 114
 circle (circumference) 110
 parallelogram 106
 rectangle 98
 square 94
 triangle 102
 trapezoid 106
Perpendicular lines 18, 104
pi 110, 112, 179
Polygon 90
Proportion 54
Protractor 21, 24
Pythagorean theorem 64, 68, 162

R

Radius 110
Ray 14
Rectangle 90, 98, 100
Rectangular solid 127, 133
Regular polygon 90

S

Sphere 179
Square 90, 94, 96
Square of a number 62
Square root 62, 63
Supplementary angles 26

T

Table of perfect squares 63
Tessellation 176
Transversal 30, 32
Trapezoid 90, 106, 108
Triangle(s) 42, 90, 102, 104
 corresponding angles 52
 corresponding sides 52
 equilateral 48
 isosceles 48
 right 49, 64, 68
 scalene 49
 similar 52, 60
 sum of angles 44

U

U.S. customary units 78

V

Vertex 14, 127
Vertex letter 20
Vertical angles 27
Volume 129, 133, 137
 cone 136
 cube 131
 cylinder 135
 rectangular solid 133
 sphere 179
 two-step problem 139, 141
 units 129

W

Work triangle 160